KUBRICK'S 2001 A SPACE ODYSSEY

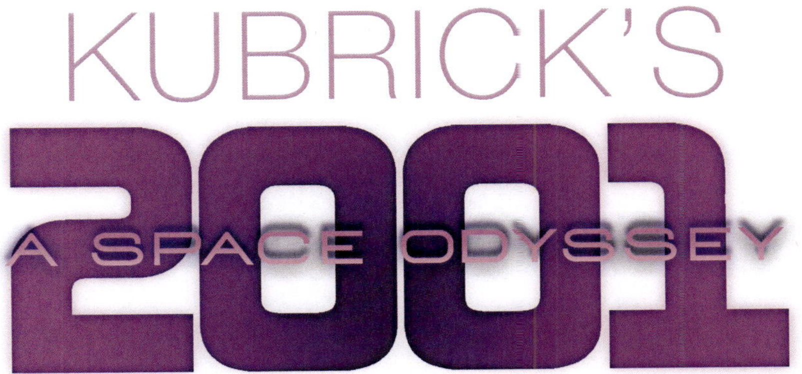

How Patterns, Archetypes and Style Inform a Narrative

second editon

ALBERT HALSTEAD

Leonine Productions, LLC

"Become yourself through what you create."

Kubrick's 2001: A Space Odyssey
2nd Edition

Cover design and artwork by Cory Freeman
Book design by Jessika Hazelton

This book is printed on acid free paper.
Printed in the United States of America
The Troy Bookmakers • Troy, New York • thetroybookmakers.com

Published by Leonine Productions, LLC
7 Spring Street
Waverly, NY 14892

ISBN 978-1-61468-428-2

for Ann Marie

Acknowledgements

The author wishes to thank the following people for reviewing the text, offering suggestions, and providing support which helped to make this book possible: Douglas Trumbull, Vincent LoBrutto, Ben Alpi, Howard Garbarsky, and Ann Marie Halstead.

Table of Contents

List of Illustrations

The illustration section is found between pages 61 and 68.

Preface to Second Edition

Our second edition arrives in conjunction with the fiftieth anniversary of the release of Stanley Kubrick's *2001: A Space Odyssey*. In celebration of this historic event, a considerable amount of new material has been added to almost every section of the book including the appendices, which now feature an exclusive interview with Visual Effects Supervisor and Film Director Douglas Trumbull. Trumbull offers us a treasure trove of new production details and insights into both his and Kubrick's work on *2001: A Space Odyssey*. All *2001* fans and scholars of the science fiction genre should find this fascinating reading.

In addition to the expanded text, more frame enlargements and photographs have been included to illustrate or provide examples of points made in the analytical essay. I have attempted to correct what I felt were ambiguities in the some of the arguments presented in the first edition and have offered more examples to reinforce many of the points made in the first edition. Also, many new references, related facts, and definitions have been added to the End Notes section in the hope that these will not only bolster our arguments, but aid Kubrick scholars and film scholars in doing further research and explain more of the film terms I have used that are likely to be unfamiliar to the lay reader.

We have gone deeper into the roots of Kubrick's mythology by more closely examining what I call his use of coherent archetypal symbolism. To support these arguments, I have quoted and frequently referred to the work of psychiatrist Carl G. Jung, particularly the volumes of his famous *Collected Works.* Accurately presenting and explaining Jung's ideas is always a daunting challenge, but I felt it provided a significant and much-needed insight into the mythological and archetypal conceptions Kubrick introduced in *2001: A Space Odyssey.*

As in the first edition, I have not referred to any of Kubrick's other films, though certainly such comparisons may offer valuable insights into the workings of *2001.* For readers who would like to learn more about Kubrick's other films and their relationship to the stylistic techniques and narrative strategies used in *2001: A Space Odyssey,* I highly recommend Thomas A. Nelson's *Kubrick – Inside a Film Artist's Maze, New & Expanded Edition,* (Bloomington: Indiana University Press, 2000). This is the best study written to date of Kubrick's entire film oeuvre and his development as an artist. It surveys all of his work, from the documentary *Day of the Fight* (1951) to Kubrick's last film, *Eyes Wide Shut,* released in 1999 shortly before his death. Particularly fascinating is Nelson's exploration of a common thread he finds in nearly all of Kubrick's work, which he refers to as Kubrick's "aesthetics of contingency."

Owing to questions and feedback we have received from a number of readers, I felt that the second edition needed to say a few words about Arthur Clarke's novel, *2001: A Space Odyssey* (published in 1968) and clarify its relationship to Kubrick's film. Long before Clarke's novel was printed, Kubrick was well aware of the differences between the story Clarke wanted to tell and the final version of the screenplay, which was largely written by Kubrick. Essentially, the two writers agreed to disagree about the story and decided that each should go his own way. Though it stands on its own as a fine piece of science fiction writing, Clarke's novel, by his own admission, does not tell the same story that Kubrick's film does. They differ in the nature of the events that occur during the story and also in the way the main characters behave. Overall, Clarke's story is not the same kind of story; therefore, it does not explain *2001: A Space Odyssey.*

As in our first edition, my belief that there is a "poetics of cinema" continues to grow and inform my writing. As Kubrick intuitively understood, a poetics of cinema has far more to do with the beauty of mathematics than it does with any argument concerning politics or sociology. My goal has been to improve and expand the original monograph for both fans of the film and those who are simply interested in learning more about how a great film works. I sincerely hope we have succeeded.

A. Halstead
January 2018

Preface

This book began as the text to a lecture first presented at the Albacon science fiction convention held in Albany, New York in October 2006. Shortly afterwards I expanded the text and considered publishing it as an article in one of the film periodicals. Eventually, what began as an essay grew into something considerably longer and I realized that publishing my work as a monograph on Kubrick's *2001: A Space Odyssey* (1968) would be more suitable.

My approach has been to examine and interpret *2001: A Space Odyssey* as an independent work of art, not as a derivative or adaptation of any other work, including Arthur C. Clarke's short story, "The Sentinel," and his novel *2001: A Space Odyssey*, published a few months after the film's release. What follows is basically an analytical essay focusing on the film's complex formal system and style. I have not attempted to relate *2001* to any of Kubrick's other films or to other films within the science fiction genre. Nor have I discussed any connection to historical events at the time of its production or ideological trends popular during the sixties. Though such approaches are certainly worthwhile and may offer additional insight, they are beyond the scope of this essay.

The first part of the essay offers a brief overview of the events in each of the film's four main sections, along with some interpretation of the most important symbols and associative elements. The essay then proceeds with an in-depth analysis beginning with the section titled "Further Analysis." From this point forward I discuss the film's formal structure, its style, and some of the various questions that have arisen regarding its themes and ideas.

2001: A Space Odyssey is not a difficult film to understand if you watch and listen to it carefully, and I recommend watching the film before reading this essay. If at all possible, view the film in a theater, preferably a 70 mm print. If a print of *2001: A Space Odyssey* is not available, the next best option would be to watch the film on a Blu-ray or regular DVD on a high-definition television set. If one of the video formats is all that is available to you, try to view the film using the best quality television set and sound system you have available.

My hope is that this essay will be enlightening not only to those who admire *2001: A Space Odyssey*, but also to those who are merely curious about it, and even to those who don't admire it. It is one of the most important films ever made and there is much that both film viewers and filmmakers can learn from it.

Any errors in the facts stated or any confusion that may result from a lack of clarity in either my arguments or my descriptions are solely the responsibility of the author.

A. Halstead
July 2009

Kubrick's *2001: A Space Odyssey*

How Patterns, Archetypes and Style Inform a Narrative

2nd Edition

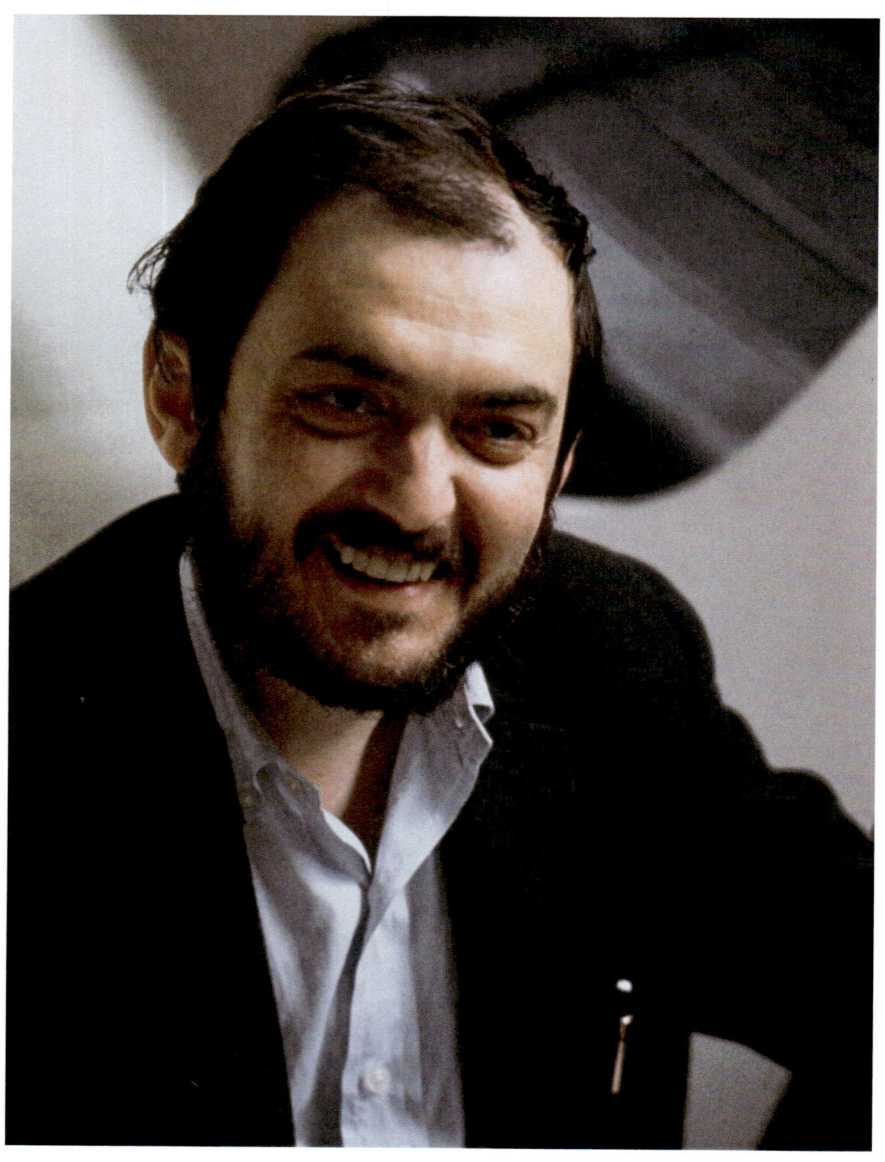

Sometimes the truth of a thing
is not so much in the think of it,
but in the feel of it.

—Stanley Kubrick
commenting on *2001*

Introduction

When Stanley Kubrick's *2001: A Space Odyssey* opened in theaters in 1968, it was immediately controversial. Simultaneously loved and hated by audiences and film critics alike, the film was hailed as a masterpiece by some while others saw it as boring and abstruse. Still others dubbed it a "brilliant failure." It even put off many who considered themselves fans of the science fiction genre. Certainly, it was a film that was largely misunderstood and, I believe, this was so because it failed to meet what audiences expected of a science fiction film in the 1960s. In a word, *2001: A Space Odyssey* is unconventional.

The key to understanding *2001: A Space Odyssey* lies in understanding how its form and style are used to convey not only a story, but three large-scale ideas involving the evolution of the human race, the evolution of technology, and encounters with extraterrestrial life. To accomplish this, Kubrick needed to expand the film beyond the conventions of the science fiction genre and the conventions of narrative film in general.

In time, with the benefit of repeated viewings and greater time to reflect, viewers and critics began to see meaning and value in *2001: A Space Odyssey's* images that they had not seen before. They became aware of not only the level of innovation and technical expertise Kubrick had demonstrated, but also the film's success in embedding archetypal symbolism in a unique form and the genius required to execute this. Many have noted that after *2001: A Space Odyssey*, science fiction films were never the same. Kubrick's film raised the bar for production values in science fiction movies and promoted the entire genre to a level of higher regard. More importantly, Kubrick transcended the genre by allowing style to play at least as important a role as story.

2001: A Space Odyssey is now regarded as one of the most important and most influential films ever made. The American Film Institute has rated it as the number one film in the science fiction genre and it ranks fifteenth on the AFI list of the "100 Greatest American Films of All Time" (2007). To this day, *2001: A Space Odyssey* remains *the* science fiction space adventure against which all other such films must be judged. This essay will, hopefully, show why as we investigate how patterns, archetypes, and style inform a narrative.

Opening Title

2001: A Space Odyssey begins with music. At first, there is nothing but blackness on the screen, but we hear, or in many cases feel through the floor, a profoundly deep organ tone. As the first three notes of the opening fanfare are played, we see the earth rising over the moon and the sun rising over the earth. These three celestial objects present a simple, powerful, majestic, and well-composed image. Centered in perfect alignment, these objects dominate the frame and immediately focus the viewer's attention. As the fanfare nears its end, the eye is drawn to the top of the alignment and the dazzling light of the sun. The alignment also forms the first grouping of three elements in the film, a significant pattern that will appear many times throughout the film. Equally important, the title demonstrates at the very outset Kubrick's command of music as a technique to amplify and harmonize with the moving image. What was needed here was a soundtrack that would command attention, accompany the visual ascent from darkness to light, and evoke a sense of triumph and awe. To this end Kubrick selected Richard Strauss' *Also Sprach Zarathustra,* the tone poem inspired by Nietzsche's account of a man's philosophical and religious awakening. This is music that knows no fear. It addresses the universe boldly and embraces it; it addresses the audience and inspires it. Notice that the ascending musical notes of the fanfare, C G C^{8va}, are repeated three times, each repetition coinciding with the ascension of a celestial body. [1] The odyssey begins with a triad in the image and a triad in the soundtrack. The film's title and, as we shall see, the triad of rising celestial bodies indicate that we are about to embark on a great journey, not only into the vastness of space, but also through eons of time and the realm of philosophical, even religious, ideas. Indeed, this image coupled with the opening *Zarathustra* theme leaves no doubt that this will be a film that is primarily about very big ideas. With the title's combination of striking visual form and music, Kubrick has set a tone for his narrative that is at once mysterious, bold, fearless, and grand in scale. He has captured the essence of the entire film within its first ninety seconds. Our adventure has begun with one of the most stunning opening titles ever put on the screen.

Act I – The Dawn of Man

The first act opens with a shot of a vast desert bathed in predawn light and the sound of moaning wind. The shots that follow present a desolate, nearly lifeless landscape, possibly some region in prehistoric Africa. A family of apes and a group of large grazing animals (tapirs) nibble on some of the sparse plant life growing in the area. One ape is attacked and nearly killed by a leopard; later two families of apes quarrel over rights to a water hole. When night falls, the apes huddle under a rock ledge watching and listening fearfully for the sounds of approaching predators. It seems they are rarely out of danger. The shots are composed in a way that allows the audience to observe their behavior from the viewpoint of an omniscient being who is far more advanced, much as extraterrestrial aliens would if they were somehow monitoring the progress of life on Earth, particularly the progress of this curious species of apes.

Although the entire first act has no voice-over narration or title cards, the images presented and the sounds we hear tell us a great deal. At this point in their evolution, the apes have not learned to use tools of any kind nor have they acquired the skill of using fire. These opening scenes clearly indicate our ancestors lived in a very hostile environment in the midst of a vast space over which they had almost no control. Every living thing has to struggle in order to survive and the apes at this point in their history seem to be barely eking out an existence. Thus, Kubrick has laid the foundation for a story that has taken the evolution of the human race as its main subject.

The apes awaken one morning to discover a large black monolith that appears to have been deliberately placed near their den. Its origin is a complete mystery and its smooth rectangular surfaces bear no inscription or markings of any kind. Owing to its precise right angles and exact rectangular form, we surmise that the monolith was fashioned by some form of intelligent being or group of beings with abilities far beyond that of the apes. The appearance of the monolith is accompanied in the soundtrack by the "Kyrie" section of Gyorgi Ligeti's *Requiem for Soprano, Mezzo Soprano, Two Mixed Choirs & Orchestra,* an eerie and intensely dramatic atonal choral work. As the apes begin to move about, they greet this strange artifact with a mixture of curiosity and intense agitation. The leader of the family cautiously reaches out and touches it, but quickly pulls his hand back uncertain and perhaps a little fearful of how the mysterious object might respond. It doesn't respond in any way that we can see or hear, yet there is something about this magnificent form that they are drawn to. Gradually, the apes gather around the base of the monolith, running their hands over its smooth dark surface, examining it with care and amazement. They seem to be aware that they are in the presence of something that is beyond them and they revere it. As the apes continue to marvel at the monolith, a striking triad pattern is formed when the sun and the crescent of the moon align in the sky directly above the monolith.

In the days that follow, the leader of the ape family (referred to as Moonwatcher in the screenplay) discovers he can use a bone as a tool. In subsequent scenes, the apes are eating fresh meat presumably from a recently killed tapir. Later on, Moonwatcher and his tribe encounter their old enemies at the water hole, but this time they are armed with bones which they use as weapons to ward off and even kill members of the other tribe.

These discoveries and primitive inventions mark a great advancement for the apes and greatly improve their chance for survival, but what role did the monolith play during these events? Could the skill of using tools been taught or somehow imparted to the apes by the monolith or is its role merely passive? The evidence available at this point allows no conclusion. However, one observation we can make is that an alignment of three things, whether three celestial objects or two celestial objects aligning with the monolith, is a pattern that has been repeated and, owing to its significance, one that we are likely to see again. Such alignments will, in fact, become a motif during the film. As the film continues, this pattern will become a subset of a more general pattern or super motif: the triad, which will recur in many forms with varying elements in both the images within the frame and the music in the soundtrack.

The importance of this pattern's meaning calls for elaboration. The triad is an archetype and an ancient symbol for the foundation or beginning of a process, including the idea of rebirth.[2] In order to understand this, we need to understand that numbers, particularly the integers below ten, each have a unique quality in addition to their significance in representing a unique quantity. [3] Imagine a space with nothing in it. Beginning with the first element, we have the quality of oneness or unity. The second element is divided from the first and stands in opposition to it, but together they create an extension–the first dimension. Adding a third element creates the triad, which establishes a second dimension or a surface that becomes a threshold to the universe or the foundation of all that will be built. So, if we were to reflect the quality of these numbers as opposed to the quantities which they represent, we would count one–unity, two–division or polarity, three–through and out into the universe.

As an archetype, the triad is found in many different cultures and societies throughout the history of the human race. Below are some relevant quotes which may offer additional insight.

> "The One engenders the Two, the Two engenders the Three, and Three engenders all things."
>
> – *The Tao Te Ch'ing*

> "The triad is the form of the completion of all things."
>
> – *Nicomachus of Gerasa*

> "Three is the formulation of all creation."
>
> – *Honoré de Balzac*[4]

Triads that occur in *2001: A Space Odyssey*, particularly those involving the monolith and celestial objects, are used repeatedly to evoke the idea of thresholds to new experiences, thresholds or doorways that opened to the human race at critical times during its evolution. The repetition of such patterns underscores their importance to the narrative, particularly the important steps in the evolutionary process the film presents. On several occasions during the film, rare alignments of the sun, planets, and satellites signify critical points in time, and times that will have a unique quality. The idea that cycles or periods of time have a quality or meaning brings this repeated pattern into correspondence with ancient astrological ideas which are also rich in archetypal symbolism and common to many cultures throughout history up to the present day.

Following his victory in reclaiming the waterhole for his family, Moonwatcher flings his bone-tool/war club high into the air. As we reach the conclusion of the first act, we have witnessed the beginning of the end of our human ancestors being dominated by their environment. They now face their world empowered by a new confidence.

Act II – A Celebration with a Shadow of Complacency[5]

A striking cut presents us with a graphic match from the bone-tool falling through the sky to a man-made satellite orbiting Earth.[6] In the blink of an eye, millions of years have passed and tools made from bones have been succeeded by satellites loaded with sophisticated, precision instrumentation constantly circling the globe. A new triad is formed by the satellite, the earth, and the sun. We hear Johann Strauss' *The Blue Danube* waltz begin to play in the soundtrack, a piece written in three-quarter time that evokes gliding, rotational movement. Next, a magnificent, rotating space station appears followed by a shot of the Pan Am shuttle *Orion* gliding towards it. The graceful movement of these objects moving through space to *The Blue Danube* welcomes us to the first year of the twenty-first century. The celebration begun by the creative ape-man of prehistoric times continues: the human race has moved beyond Earth into space and our tools have evolved with us to the point where they may now operate without us, and millions of miles beyond our direct physical control.

Inside the *Orion*, a stewardess in zero gravity cautiously treads her way down the aisle to attend to a passenger. In the shuttle's cockpit, we see the pilot and copilot at the control console. This shot provides an excellent example of Kubrick's use of cinematography and mise-en-scene working together as stylistic techniques to convey what is most important about the scene.[7] The camera is placed so that we are viewing the console from a position between and behind the pilots. Consequently, we see them seeing what we are seeing–the *Orion's* control console with its graphic displays, telemetry, and computer-assisted navigation. This is the subject of the shot, not the pilots. As the subsequent tighter shot confirms, this is what Kubrick wants us to see–machines that have advanced to the point where they are able to detect and communicate important information: a significant point in the evolution of the tools that humans depend on and, as we shall see, a prelude to artificial intelligence.

Back in the passenger compartment, we find our second protagonist and modern-day successor to the tool-inventing ape, Dr. Heywood Floyd. He appears to be the only passenger on board, and we see that this incredible journey (especially for audiences in 1968) and the technology which surrounds him have become so familiar and mundane that he has fallen asleep during the flight. Contrast this to the curiosity, amazement, and jubilation of the ape when he discovered that a bone could be used as a tool.

The *Orion* synchronizes its rotation with the space station to prepare for docking and the two machines dance in synchronous harmony. We continue on a journey through space and time witnessing all that the human race has accomplished: a grand, elegant clockwork where people appear insignificant compared to their machines and have been nearly assimilated by them. On arriving at the space station, Dr. Floyd engages in banal conversation with the station's staff and telephones home to wish his daughter a happy birthday. All seems uneventful and routine until he meets an old friend, Elena, who is relaxing with a group of three other Russian scientists who have just come from the moon and are returning to Earth. The conversation starts out friendly and relaxed, but becomes increasingly tense when one of the Russians, Andrei Smyslov, questions Floyd about odd things that have been happening at an American moon base near Clavius. Smyslov presses Floyd for some indication of what the nature of the problem might be and inquires about a report that a "serious epidemic has broken out at Clavius." Floyd remains tight-lipped however, and tells him he is not at liberty to discuss it. The audience is now confronted with a second mystery, the first being the source and function of the monolith that the apes encountered earlier.

Heywood Floyd's voyage continues aboard a second transport, the *Aries*, which takes him to the US moon base. Like the *Orion*, the *Aries* approaches its destination maintaining a graceful pace with the music and we encounter another triad formed by the *Aries*, the sun, and Earth. We also have another shot of the pilots at the control console completely absorbed in the constant stream of information it is feeding them along with its shifting patterns of color. This kind of shot is another motif that will be repeated many times in the film. As the *Aries* approaches the landing site, Kubrick allows sufficient time for us to take in the extraordinary detail of the architecture and the complex operation of both the landing facility and the descending transport. We wonder at a technology so advanced and grand in scale that it would have appeared magical to humans a century before, not to mention the ape-men. The magnificent harmony with which the machines of the twenty-first century function, reinforced by the music in the soundtrack, constantly conveys the idea that this is a time of celebration for technology and an abundance of confidence for the human race. However, it is equally interesting to note that the importance of the humans at this point in their history appears to be waning. Their behavior and dialogue are neither interesting nor dramatic. Heywood Floyd continues to sleep through most of the journey and we have learned more about the operation of the machines he and his contemporaries live with than we have learned about them as individuals. We also observe that the interiors of these twenty-first century machines, though very clean and well- ordered, tend to be somewhat stark and empty, lacking a lived-in look. The humans seem awkward and rather out of place in such surroundings. Needing food and toilets, they are almost a form of pollution living inside the machines. You get the impression that if these machines could think, they would consider themselves superior to the human race.

After arriving at the lunar base, Floyd attends a briefing where he is introduced by Dr. Ralph Halvorsen as a spokesman for the National Council on Astronautics. Floyd then addresses the administrators and his fellow scientists, apologizing for any inconvenience or distress the cover story concerning the mysterious events at Clavius may have caused. We eventually learn that the great secret the Council on Astronautics and the scientists at Clavius have been guarding has nothing to do with any sort of epidemic, but concerns instead the discovery of something which, in Floyd's words, "may well prove to be among the most significant in the history of science." Floyd expresses his appreciation for the patience and forbearance his colleagues have shown, but insists that the policy of tight security and "absolute secrecy" must continue. Both the Council on Astronautics and the American government believe the shock this discovery might create among the public could set off a worldwide panic. When Floyd finishes his talk, we haven't learned what the nature of the discovery is, but Floyd reveals that his mission during this visit is to gather information and subsequently prepare a report for the Council that will contain his recommendations as to "when and how the information on the discovery should be released."

Notably, both Floyd's speech and his demeanor appear overconfident and too casual, particularly when we consider the seriousness of the national and global impact to which he refers. When Floyd asks if there are any questions, he receives only one from a colleague concerned about how long the phony cover story will remain in effect, while the rest of his audience remains mute. In fact, Floyd and everyone else in the room seem to be remarkably devoid of any excitement or curiosity about the amazing discovery. As strange as it may seem, this peculiar character behavior represents that of most of the characters we have seen thus far, and it indicates that their role and Floyd's role, in particular, are part of a larger scheme where events unfolding on a broader scale matter more than the activities of any single character.

Floyd, Halvorsen, and several other scientists from the base take a lunar transport to the site where the "discovery" was made. During the flight, they eat synthetic sandwiches and congratulate each other on the job that each has done in investigating what was discovered and maintaining the phony cover story. Floyd is shown several photographs of the site as his companions explain how the detection of a powerful magnetic field emitted by an unknown object lying several feet beneath the surface led them to discover what they now believe is some kind of artifact left on the moon by extraterrestrial beings. Halvorsen adds that they have determined that the artifact, labeled TMA-1[8] in the photographs, was "deliberately buried" beneath the lunar surface some four million years ago. All too self-satisfied, Floyd chuckles and shakes his head upon hearing this, then concludes the conversation with an extraordinary bit of understatement, casually responding, "Well, I must say, you guys have certainly come up with something."

As the transport passes over the moon's surface, we hear strains of Ligeti's ethereal choral piece, *Lux Aeterna,* and observe the stark yet beautiful lunar landscapes. A triad formed by the transport, the earth, and moon heralds the coming of another important moment for the human race which is not only on the threshold of exploring space beyond the moon, but also about to deal with the first evidence of extraterrestrial life. The unearthly, Gothic lunar landscape resonating with Ligeti's music imparts a mysterious, religious quality to the scene, a quality completely at odds with the scene we witnessed inside the lunar transport.

Floyd and company arrive at the excavation site and we see that the artifact that has been unearthed is actually a second monolith that looks identical to the one discovered by the apes. The volume of the Ligeti *Requiem* in the soundtrack rises as the group approaches the monolith. The chorus in *Requiem* seems to be anticipating some great revelation as Floyd walks to the monolith, then reaches out and touches it. The ancient artifact seems to emit an audible hum, and, for a moment, we wonder if Heywood Floyd has any sense of how far he is reaching back into his prehistoric past, making contact with this extraterrestrial artifact just as his simian ancestors did. Such ideas are quickly dispelled, however, as the group gathers for a photograph, demonstrating more interest in making a record of themselves at a historic discovery than the discovery itself.

Suddenly, we hear a piercing, high-pitched sound coinciding with a second extraordinary alignment formed by the monolith, with the sun and Earth rising directly above it. Our protagonists are again standing before a threshold about to take another giant step in their journey. Yet, what is most peculiar and prominent in the scenes leading to the end of Act II–Floyd's speech at the moon base, the conversation in the moonbus, and the group's behavior at the excavation site–is that in spite of this astounding discovery, the mystery of its origin, and the import of the rare celestial alignment occurring simultaneously with the high-pitched radio signal, Heywood

Floyd and his colleagues have exhibited remarkably little curiosity and none of the awe or wonder that one would expect to see in response to these extraordinary events. What we witness instead is a disturbing amount of vanity, self-assuredness, and complacency. As we leave the second act, Kubrick has given us a sense that the human race is approaching an event that may forever alter its consciousness and that the alteration is something for which the humans in this chapter of their history are totally unprepared.

Act III Part I – Have Humans Gone Too Far?

Our journey continues with a shot of the interplanetary transport, *Discovery One*, moving slowly through the blackness of space en route to Jupiter. Eighteen months have passed and humans have now ventured farther from Earth than ever before. We are introduced to two new characters, Frank Poole and David Bowman. As they exercise and calmly go about their daily routines, the music in the soundtrack evokes an atmosphere that is very different from the celebration that opened Act II. Initially, the mood here is one of reservation, even the ship's name reflects a less aggressive stance than the names of its predecessors, the *Orion* and the *Aries.* The subtle, somber tones of the "Adagio" from Aram Khachaturian's *Gayane Suite* gradually evoke feelings of sadness and foreshadow that something ominous and disturbing may take place on board the *Discovery.*

As Dave Bowman and Frank Poole settle in for their evening meal, they watch a television news program being broadcast from Earth. In a previously recorded interview, a reporter named Aimer interviews Dave and Frank about the *Discovery's* mission and life aboard the ship. He then proceeds to interview the *Discovery's* onboard computer, a HAL 9000, which he has learned "one addresses as Hal."

We learn that Hal possesses an extraordinary degree of artificial intelligence and actually runs the entire ship, leaving Dave and Frank with little more to do than monitor the *Discovery's* progress. Hal is trusted to take care of everything within the ship, including preserving the lives of the three crew members who are in a state of hibernation. When Aimer asks Hal if he has experienced any lack of confidence owing to the enormous responsibilities placed upon him, Hal responds, "Let me put it this way, Mr. Aimer, the 9000 series is the most reliable computer ever made. No 9000 computer has ever made a mistake or distorted information. We are all, by any practical definition of the words, foolproof and incapable of error." Addressing Dave Bowman, Aimer points out that Hal seems to show more than a little pride when he speaks about the 9000 series' perfect record. He then asks Dave if Hal's emotions are genuine. Dave responds that Hal was programmed to simulate human emotions because it would make it easier for the crew to interact with him, but as to whether or not Hal's emotions are real, Dave doesn't believe anyone really knows.

In subsequent scenes, Hal maintains an extremely polite and considerate manner towards Frank and Dave. His attention to their every need is almost motherly. He responds with more emotion than Frank after Frank's parents celebrate his birthday in a video transmission. At times, Hal's sensitivity and emotional responses to his crewmates seem to make him more human than his human crewmates. With Hal, we see that machines have now evolved to the point where their technical perfection combined with their ability to respond emotionally has seduced the humans who work with them into trusting them to perform a critical role in a long, complex mission that includes piloting the ship and monitoring life support for all five astronauts on board. With the advent of Hal, another milestone has been passed: both intellectually and emotionally, our human protagonists now stand very much in the shadow of their machines. This situation is affectingly brought to the fore in the scene where Frank watches the video from his parents, and again, in the scene where Hal defeats Frank in a game of chess. Frank is humiliated, though Hal covers for him by being very gracious. Throughout these two scenes, Kubrick uses Khachaturian's "Adagio" to evoke a funereal atmosphere, reminding us of the eclipse of humanity that is taking place while Frank and Dave show no awareness of it.

As we approach the end of the first part of Act III, Hal lures Dave into a conversation by expressing an interest in Dave's sketches. Hal then moves on to the matter he is really concerned about by asking Dave if he has had any second thoughts about the mission. Hal probes Dave to see what Dave might know about strange stories and rumors going around before the start of the mission. Dave listens carefully, but either doesn't know or can't say what he knows about these stories. Hal apologizes for being "silly" in asking such questions. Dave maintains a poker face, but in his eyes we can sense that he is wondering, as are we, why such a peculiar line of questions? What is Hal really trying to find out and why is he so concerned about the mission? Dave has witnessed the first sign of a crack in Hal's cybernetic personality and now it is Hal who feels embarrassed, as he has tipped his hand in his unsuccessful attempt to manipulate Dave.

Could the red color of Hal's omnipresent, electronic eye betray a malevolence that has yet to surface? Has artificial intelligence advanced to the point where machines are ready to act on their assumption that they are superior to human beings?

Act III Part 2 – Our Tools Betray Us

As Hal is about to conclude his conversation with Dave Bowman, he suddenly detects a fault in the ship's radio antenna circuit, referred to as the AE-35 unit, and insists it will fail in seventy-two hours. Dave retrieves the unit and tests it, but can't find anything wrong with it. Hal's twin HAL 9000 on Earth analyzes the unit, but disagrees with Hal's forecast. When Dave asks Hal to explain the discrepancy, Hal answers stating quite confidently, "...it can only be attributable to human error."

Neither Frank nor Dave shares Hal's confidence, though they try to avoid making Hal aware of their doubts. To share his concerns with Frank privately, Dave initiates a ruse asking Frank if he could help him fix a problem with the transmitter in one of the space pods. Dave and Frank enter the pod, seal its door, and check to see if Hal can hear them speaking. Once they are satisfied that he can't, they discuss the problem of whether or not they can rely on Hal to function normally for the remainder of the mission. They agree that if the AE-35 does not fail then the problem lies with Hal and he will have to be shut down. This would involve disconnecting all of Hal's higher brain functions, but then a troubling thought occurs to Dave: "Well, as far as I know, no 9000 computer has ever been disconnected." Frank: "No 9000 has ever fouled up before." Dave: "That's not what I mean." Frank: "Hmm?" Dave: "Well, I'm not sure what he'd think about it."

Unfortunately, by observing Dave and Frank's conversation through the space pod's window, Hal is able to read their lips. Hal discovers their plan to disconnect him and proceeds to kill Frank while Frank is outside the *Discovery* preparing to replace the AE-35 unit. Dave immediately sets out to retrieve Frank's body, but after he has left the ship Hal murders the three hibernating astronauts by cutting off their life support as they lie helpless in their hibernation modules. At this point, the evolution of man's tools has taken yet another major turning point. We now have our first case of machine turning against humans, in this case plotting and deliberately killing four people. When Dave returns, Hal refuses to let him enter the pod bay and arrogantly declares, "This mission is too important for me to allow you to jeopardize it." However, Dave outwits Hal and, at the risk of losing his life, manages to reenter *Discovery* through the ship's emergency air lock. From the air lock, Dave enters the pod bay still wearing his space suit and a helmet to insure that he will have a sufficient air supply without relying on Hal. Hal pleads with Dave and tries to persuade him that he is now functioning normally, but Dave ignores him and proceeds to the Main Logic Center where he will disconnect Hal's vital circuits to shut him down. Man must now kill his tool in order to survive. Hal dies singing the lyrics to the nineteenth-century love song, "Daisy Bell," in one of the most ironic scenes to ever appear in a science fiction film. Most importantly, humans have now regained center stage in the drama.

Immediately after disconnecting Hal, Dave hears the voice of Heywood Floyd coming from a pre-recorded video message that had been stored on board before the *Discovery's* mission had begun. Floyd's message explains that the true purpose of the mission was made known only to Hal and that it is connected with the monolith discovered on the moon. Dave learns that the crew was to investigate the receiving end of a radio signal being sent towards Jupiter by the monolith and that this was the only thing the scientists were able to learn about the monolith.

Act IV – Jupiter and Beyond the Infinite

The *Discovery* has now reached an orbit around Jupiter at an altitude roughly equivalent to that of some of Jupiter's larger moons. Dave leaves the ship using the remaining space pod and sets out, presumably, to investigate exactly where the radio signal is being sent. He discovers what appears to be a huge monolith floating in orbit around Jupiter. A triad is formed by the monolith, the sun, and Jupiter as the siren-voices of Ligeti's *Requiem* beckon us to a new threshold. A series of spectacular shots show Jupiter's moons and the monolith coming into alignment with the sun. Having encountered such alignments earlier in our journey, we know that this is likely to signify a critical time and that something extraordinary is about to happen. Dave approaches the monolith, but discovers that what really lies before him is a portal to what appears to be a great, black abyss. He crosses this threshold as the chorus in Ligeti's *Requiem* rises once again and we share Dave's point of view as he embarks on a journey through space and time that the aliens now control. Dave is helpless and terrified, and we have lost the advantage of our omniscient point of view. We have no idea where Dave is going and are both dazzled and intrigued by the patterns of light and color through which he is flying.

Although this sequence has a narrative function–Dave Bowman's transportation to the alien world–its originality and artistic significance become more important when viewed as an abstract sequence within the film's plot,[9] where what is most important is the relationship between the images, their patterns of color, the music, and the sensations these evoke. Kubrick has tried to give us an impression of what travel to a distant galaxy at superluminal speeds would feel like. Here, the space-time continuum has been warped or folded and we travel by means beyond spacecraft, or anything we understand about physics as we know it today. Kubrick correctly understood that whatever we would see or sense would be unlike anything we had ever experienced before. At times Dave seems to be traveling back in time, perhaps to the dawn of creation itself, witnessing the birth of whole galaxies and star systems. Five diamonds appear ahead, then seven as Dave approaches primordial landscapes rephotographed using contrasting pairs of primary and complementary colors. These shots are intercut with extreme close-ups of his eye using the same solarization effect in the photography and similar complementary colors.

The star gate sequence ends abruptly with a cut from an extreme close-up of Dave's eye (returned to natural color) to his point of view looking out through the space pod's window. At first Dave is inside the pod, but then he is outside. To his amazement, the pod has landed in an elegantly furnished bedroom and although he is alone in this place, it appears the aliens have tried to construct a setting in which he will feel comfortable. The decor and furnishings are from an earlier century when one did not have to worry about machines that could think, deceive, or murder. Yet, in spite of their luxury and familiarity to any visitor from Earth, Dave's accommodations could also be viewed as a laboratory or cage in which he could be studied.

We hear strange-sounding voices, presumably the aliens', but we never see them.[10] We hear Dave's breathing and notice that he has aged considerably. Dave walks about the room observing the adjacent bathroom, his reflection in the mirror, and his aged self eating at a table. He is as baffled by all of this as the apes would have been trying to understand their reflection in a mirror. Apparently, the rules of Newtonian physics with events happening linearly in space-time do not always apply here. Dave tips over his glass of wine and it shatters as it strikes the floor. He stares down at the shattered glass then raises his gaze. In the next shot, he sees himself lying on the bed, near death. Both shots represent destruction, decay, and the inevitable increase in entropy[11] that accompanies such processes. The second law of thermodynamics[12] does operate in this part of the universe, [13] yet from a holistic point of view, Dave's death would signify a new beginning or even a renaissance. Making its final appearance in the film, the monolith stands at the foot of the bed. Dave raises his arm, pointing to it, then the shot is reversed and we see a transparent capsule on the bed emitting an almost blinding light. Inside the capsule lies an infant: Dave has transcended his old body and become reborn.

The final scene: To the opening C G C*8va* notes of the *Zarathustra* fanfare, we come full circle back to Earth with the Star-Child in orbit. The final triad is formed as the frame descends from the moon to the earth and Star-Child. Through this triad, Dave Bowman's journey is complete. Like the apes before him, he has survived a hostile environment in a vast space and evolved into a new form of being. He becomes the final element in yet another super triad created by the periods of time that have unfolded during the film with the earth corresponding to the past, the moon the present, and the Star-Child the future. Whatever the purpose of the Star-Child may be, Kubrick's odyssey of the human race ends on a hopeful note expressed in the triad, the elements that compose it, and the *Zarathustra* theme. The aliens have reached out to us with an unborn child. As the camera moves in, we see an expression that is mild and innocent. With eyes wide open with wonder, the Star-Child looks down upon the earth. Kubrick has given us a universal image of hope. The three notes of the fanfare, together with all of the aforementioned elements of style, have opened a new door for us and another journey is about to begin.

Further Analysis

Whether he understands them or not, man must remain conscious of the world of the archetypes, because in it he is still part of Nature and is connected with his own roots.

– *Carl G. Jung*

Kubrick's Multi-Formal Approach

Through the art of cinema, Kubrick has expressed three major ideas in *2001: A Space Odyssey*: 1) the evolution of man's tools, machines, and technology is an awesome thing to behold with consequences that may not be to our benefit; 2) the exploration of space and encounters with alien life are mind-boggling experiences; and 3) the evolution of the human race and its destiny are awesome things to contemplate. These are ideas that are certainly great in both scale and scope, requiring a story spanning millions of years. Kubrick realized that presenting such ideas through a conventional narrative film would have overloaded the structure and required far too much screen time. His genius as a director lies in his solution to this problem. Rather than relying on forms of text (as in written or spoken language) or emotional character behavior, Kubrick took the bold step of relying on the power of imagery, particularly symbolic imagery and imagery driven by music, to express his ideas. To this end, he integrated strategies from two forms of non-narrative film, specifically abstract and associational film form, into what is essentially a classical narrative film.

To understand the advantage of this multi-formal approach, we need to understand first how a film–any film–works. The shadows projected on a screen that make up what we call a film or movie are organized for some purpose. A film's formal system is what gives direction and purpose to the images moving on the screen.[14] Narrative form organizes images to tell a story that involves characters moving through a literary plot, i.e., a plan, involving a sequence of cause-and-effect related events usually reaching some goal or logical conclusion. Nearly all of the films that we view at the local cineplex or on television follow this form. An abstract formal system organizes images primarily "through repetition and variation of such visual qualities as shape, color, rhythm, and direction of movement."[15] Examples of some famous abstract films are *Ballet méchanique* (1924) directed by Léger and Murphy, *Symphonie Diagonale* (1925) by Viking Eggeling, *Parabola* (1936) by Mary Ellen Bute, *Motion Painting #1* (1947) by Oskar Fischinger, *Mothlight* (1963) by Stan Brakhage, and *Serene Velocity* (1970) directed by Ernie Gehr.

An associational film organizes images "to suggest similarities, contrasts, concepts, emotions, and expressive qualities."[16] Some important examples would be *Man with a Movie Camera* (1929) directed by Dziga Vertov, *A Movie* (1958) by Bruce Conner, *Unsere Afrikareise* (1965) by Peter Kubelka, and *Naqoyqatsi* (2002) directed by Godfrey Reggio. Abstract and associational films may suggest ideas and/or evoke emotions through their graphic, pictorial, or symbolic qualities, or through editing that allows us to see some similarity in meaning or concept linking a set of images. Such films do not rely on characters or follow a literary plot.

Abstract form predominates in two sequences in *2001: A Space Odyssey*, the opening title shot and the star gate sequence, and it becomes quite prominent as it commingles with narrative form in the waltz sequence that opens Act II. In fact, this sequence has been hailed by several writers as a true "Ballet méchanique" in reference to Murphy and Léger's pioneering film of the same name. Kubrick's use of abstract film strategies continues throughout *2001*, sometimes in a striking way as in those scenes where planetary alignments appear, or more often, in a subtle, pictorial way operating just beneath the plot's narrative surface as in the frequent close shots of control consoles or instrument panels, the scenes involving spacecraft in flight, and shots containing non-aligned triads. Kubrick used abstract form to stimulate our sense of sight and sound directly, relying on composition, visual aesthetics, and choreography supported by and integrated with music in the soundtrack to captivate our attention and convey the emotional tone of a scene. In the opening title for example, the general idea the formal qualities convey is a powerful awakening, rising, and moving towards the light. During Dave Bowman's flight through the star gate, extreme close-up shots of his eye are intercut with patterns of light and color that he is moving through. But the eye in such an extremely tight shot does not reveal anything about the subject's emotional state or indicate what he might be thinking. Instead, the unnatural color of the iris reflects the unnatural and traumatic journey Dave is experiencing while the large dark pupil presents us with yet another mystery as inscrutable as the blackness of deep space through which he is speeding. Kubrick uses this altered image of the eye to help us understand that being transported a distance incomprehensible to twentieth-century humans at a speed beyond that of light is altering Dave's consciousness.

Kubrick's associational strategies become apparent as we make the transition from the opening title to the first act, where we see the correspondence between the rising sun and the light of dawn in the shot that begins the "Dawn of Man" section. It is quite fitting that Kubrick chose a desert as the setting for humanity's beginning. The desolate wilderness has been, since biblical times, a furnace where prophets and holy men go to be tested, forged, and re-formed. In its emptiness lies its vast potential; it is the creative void and a place where both the blessed and the ordinary often experience enlightenment or an epiphany.

A second associational transition occurs between Act I and Act II with the famous cut from the bone-tool to the satellite. The two images are juxtaposed so that we may contemplate the evolution of man's tools, the evolution of the human brain that conceived them, and the passage of millions of years of human history seeming like only a moment to a highly advanced extraterrestrial race.

In retrospect, a number of shots in Act III now appear to offer far more meaning than we might have sensed initially. We have the long, side shot of *Discovery* looking like the sun-bleached skeleton of a dead fish drifting in a dead sea; then a human running in circles, going nowhere, passing three of his crewmates who are seemingly lifeless and already lying in their coffins. Meanwhile, the omnipresent eye of the machine takes all of this in–a detached observer, silently biding its time and jealously guarding its secret. We begin to feel as if we were watching a scene in which someone was seriously ill and dying, but these images which evoke associations with moribund subjects integrated with the pathos of Khachaturian's "Adagio" represent more than the fate of Frank Poole and the three sleeping astronauts. They are a metaphor for a fading human spirit; a spirit that once had passion, curiosity, and verve that fueled a desire to explore, discover, and invent.

As Bowman approaches the alien world in the beginning of Act IV, a row of five diamonds appears in the sky ahead, increasing to seven. Diamonds have been long associated with perfection, but the numbers five and seven have significant archetypal qualities. Five, from which the pentagram and the golden section are derived, has an ancient association with regeneration and life. Seven represents the steps to perfection, the unattainable, and self-transformation. Thus we have a strong connection between these elements and the fate about to befall Bowman in the alien suite.

The final sequence in the alien suite offers a garden of opportunities to form associations such as Bowman's observing himself in the mirror, then himself seated at a table, eating a meal. The first is a reflection in space, the second a reflection in time which, through a kind of out-of-body experience, affords Bowman a unique opportunity to observe objectively his accelerated aging and contemplate the final step in his journey. This leads to Bowman's elegant last supper, which suggests a ritual of preparation for his death and resurrection, and finally, Bowman's contemplation of the shattered glass of wine, which represents the inevitable outcome of an irreversible process, not unlike his very temporal life, his deterioration, and his death. From the beginning of Act IV to end of the film, Kubrick stimulates our imagination to really take flight. Through an act of cinematic poetry, he allows us to experience something beyond anything we have seen before. Thus we do not merely watch Bowman's bewildering and astounding experience, we share it with him.

Other prominent elements rich in associations include the monolith which suggests a doorway, threshold, or opportunity; and the rare alignments which suggest creative time, major significant events at that time, evolution's relationship to time, the relationship between meaning and time, and synchronicity. Symbolism at a more general level occurs with the appearance of the various triads that suggest foundations, thresholds, regeneration, and the attainment of greater wholeness. The list of associations could be extended to finer details within the sets and staging, such as the blackness of the monolith and outer space representing deep, inscrutable mystery, nothing and everything in one, the infinite. Much of the pleasure of watching an associational film derives from the freedom one has to make associations. The preceding examples are merely my interpretations. As with poetry, there are no right answers. Through either their formal qualities or the concepts and emotions we may infer from their association, we derive meaning from the images, the importance of the meaning being proportionate to the recurrence of similar patterns (motifs). The film's motifs all have formal qualities which grab the viewer's attention and reinforce the idea of similarity, but it is primarily their associational qualities interacting at key points with Kubrick's allegorical narrative that make the most meaningful impressions on our mind.

The film's narrative form also relies on repetition in visual forms and key elements of mise-en-scene, as can be seen in the use of similar patterns in the staging of characters' activities and events to allow parallels to be seen between scenes in different sections of the film. The scene where the apes discover the monolith, for example, and gather around it while the rising sun forms an alignment with the monolith and the moon is paralleled by the scene at the end of Act II when Heywood Floyd and his group gather around the monolith unearthed on the moon. Again, a similar alignment forms while the same piece of Ligeti's music plays in the soundtrack.

In general, story information conveyed by dialogue in *2001* is minimal, occurring in less than forty-one minutes of the film's 139-minute running time. Lengthy, less-stimulating, expository dialogue along with voice-over narration, both often used in narrative films covering long spans of time, have been completely avoided. Titles appear occasionally, but these are quite brief and used only to establish the time or setting of some of the acts. As we have seen, Acts I, IV, and many of the sequences in Acts II and III rely entirely on the visual image to convey narrative information. Music and lighting, along with patterns of color and motion, are used most often to convey the emotion of the various scenes while emotional speech, facial expressions, and gestures are generally minimized. Albeit unconventional, Kubrick's complex formal strategy yields a film with remarkable economy in its use of text without sacrificing communication. This, in turn, brings the audience closer to a first-person experience as it watches the film, which is exactly what Kubrick was striving to achieve.

Story and Mythology

2001: A Space Odyssey was criticized for having a disjointed story plot, but its plot unfolds quite linearly in a structure that is not atypical of classical Hollywood cinema. The story plot is composed of four major sections[17] that trace the journey of the human race as it evolves over time and begins to expand out into the solar system. Beginning with the "Dawn of Man" section[18] set in a time roughly four million years ago, we see the human race take its first humble steps. We then make a spectacular leap to the year 2001, when humans and their machines have become far more sophisticated and begun to explore space. In the third act, we move on to the Jupiter mission and the resolution of the inevitable conflict between humans and artificial intelligence. For the final act, we travel beyond time, concluding the odyssey with the birth of the Star-Child and his return to Earth. Below is a more detailed breakdown which demonstrates some of the many parallels that occur among the various sequences.

- Prehistoric apes eke out an existence in a hostile environment that controls them.
- A mysterious monolith appears, origin and purpose unknown.
- The ape-men discover how to use primitive tools and begin to take control of their environment.
- We leap ahead four million years; man rules the earth and moon, yet his tools have also evolved and are on the verge of outshining him in their ability and splendor.
- A second monolith is discovered on the moon and believed to be alien in origin, its purpose also unknown.

- Humans travel to Jupiter, purpose of mission not known at first. While their vitality continues to languish, a new advanced machine on board can think, respond emotionally, and is self-aware.
- Machine becomes paranoid and nearly succeeds in killing all of the human crew.
- Human, alone in a hostile environment, terminates machine, regains control, and continues with a new mission and a new awareness.
- Man is studied by the aliens and re-encounters the monolith. The time has come for him to die and be reborn with a new set of abilities and a new mission.
- The Star-Child returns to Earth and a new era is about to begin.

We have seen how symbolic archetypes such as the triad have informed many scenes during *2001*, but it is also evident that Kubrick employed such mythological archetypes[19] as the hero of the Hero's Journey, the shadow personality, and the child-god or child-savior, particularly during the last two acts. In his book *The Hero with a Thousand Faces*, author and comparative mythologist Joseph Campbell describes the common characteristics found in the Hero's Journey type of story or myth, which includes the story's protagonist, the mythological hero, and his/her antagonist, the Shadow. Below is Campbell's outline of the Hero's Journey:

> The mythological hero, setting forth from his common day hut or castle, is lured, carried away or voluntarily proceeds, to the threshold of adventure. There he encounters a shadow presence that guards the passage. The hero may defeat or con-

> ciliate this power and go alive into the kingdom of the dark… Beyond the threshold, then, the hero journeys through a world of unfamiliar but strangely intimate forces some of which severely threaten him (tests), some of which give magical aid… The final work is that of the return. If the powers have blessed the hero he now sets forth under their protection (emissary); if not he flees and is pursued… At the return threshold the transcendental powers must remain behind; the hero re-emerges from the kingdom of dread (return, resurrection). The boon that he brings restores the world (elixir).[20]

Dave Bowman's adventure on board the *Discovery* and in the alien world closely parallels the journey of the mythological hero Campbell has described. Bowman must pass through a number of thresholds and survive a number of threats and tests to complete his journey or life mission. Hal's role may have been the "shadow presence that guards the passage." He is certainly the primary agent in the first test that Dave must overcome. In the face of Hal's instability, Dave exposes himself to great risk when he ventures out to recover Frank's body and then again when Hal refuses to let him enter *Discovery* through the pod bay door and he is forced to enter through the vacuum of the emergency airlock without a space helmet. Dave is put through further ordeals by the aliens as they transport him to a distant galaxy. The voyage itself seems to have put him into a state of shock. Then he experiences an intimate and profound disorientation as he adjusts to the otherworldliness of the alien suite. Finally, Dave's aging process is accelerated, which eventually leads to his death. Yet, in the end, the aliens or some other unknown power have "magically aided" his rebirth and return to Earth as the Star-Child.

By the end of the film, we understand that Dave Bowman's role is more argonaut than astronaut, yet he really represents all of us as an "everyman" bearing witness to the close of the second millennium while advancing to the next. His is the penultimate part in a drama that has presented the human race as the main character throughout the film. Several different characters have represented the human race in the course of *2001's* plot. Beginning with the ape-men who represented the human race when it was in its very primitive and humble infancy, we then leaped millions of years ahead, where Heywood Floyd and his contemporaries represented humans at the end of the twentieth century. A year and a half later, machines have evolved to the point where they eclipse the new representatives of the human race, Dave Bowman and Frank Poole. Poole is slain, but Bowman survives to vanquish the machine and initiate the revival of the human race.

One might wonder why the role of the protagonist is shared by two characters in Act III. What is Poole's role? During the process of evolution, we know that some species fail to survive. Frank Poole represents not so much a kind of human being, but a way of thinking and behaving that is destined to become extinct. In retrospect, his death is inevitable and we begin to become aware of this as we witness his detachment from his human roots, as emphasized in the scene where he watches the video transmission from his parents. Moreover, Poole's thinking is too linear and lacks creativity. He struggles to keep abreast of the machine, but in the end it overwhelms him and drains the life out of him.

Though Poole and Bowman resemble each other on the surface, Bowman's personality is the complementary opposite of Poole's in a number of significant ways. He is more thoughtful, open, and able to adapt. Bowman observes, listens carefully, and considers the consequences of his actions. He does not challenge Hal; he quietly engages in activities that Hal can't do. By the close of Act III, Bowman has emerged as the more forward-thinking of the two and the one best equipped to become the "mythological hero."

Hal's role is perhaps the most pivotal of any character in *2001: A Space Odyssey*'s parallel tales of machine and human evolution. When we first meet Hal, he appears to represent the summit of human technological progress, a creation made in man's image, yet incapable of error. He serves as the antagonist to Frank Poole and, particularly, to Dave Bowman, and, just as importantly in Kubrick's mythology, Hal becomes an example of the machine or human creation that went too far for its own good, ultimately challenging its creator for supremacy. However, in completing Hal's character and his role, Kubrick has given him qualities that are reminiscent of what Jung called the "Shadow" archetype or "negative side" of human personality:

> "... the Shadow, that hidden, repressed, for the most part inferior, and guilt-ridden personality whose ultimate ramifications reach back into the realm of our animal ancestors and so compromise the whole historical aspect of the unconscious."[21]

More than acting as the Shadow of Dave Bowman, I believe Kubrick saw Hal as the Shadow of the whole human race. Hal's fears and mistrust of his crewmates harken back to the apes' fear of predators and losing the use of their water hole, not to mention Heywood Floyd and the Council's need for tight security and "absolute secrecy" concerning the discovery near Clavius, and even Smyslov's anxiety over the nature of the mystery at Clavius and Floyd's refusal to discuss it. However, Hal proves to be much less comfortable with deception and secrecy than Heywood Floyd. His brain/programming cannot resolve this kind of inconsistent behavior and as we move through the third act, Hal's Shadow qualities begin to emerge. His growing discomfort over having to deceive his crewmates about the true purpose of the mission combined with his pride and arrogance eventually develops into full blown paranoia which in turn leads Hal to devolve into a being whose behavior increasingly resembles that of the leopard or the killer/shadow side of the apes of the prehistoric desert more than either his human creators or his human crewmates aboard the *Discovery*.

The final form of the mythological protagonist makes its appearance at the very end of the film as the Star-Child, who appears to be derived from the mythological archetype known as the child-god or child-savior. Jung has described this archetype as a unique being who "frequently appears through a miraculous birth" and is gifted with the ability to "heal or make people whole." [22] In sacred texts and artwork, he or she is often depicted within a circle or sphere. Given the role this archetypal character usually plays, the inspirational quality of the *Zarathustra* theme heard in the soundtrack, and the beauty of the images, it would seem most likely that the Star-Child's mission will be to somehow alter the course of the human race for the better.

Analysis of *2001: A Space Odyssey*'s Style

If Kubrick's use of multiple formal strategies in one film was a stroke of genius, his way of integrating the stylistic techniques was equally original and brilliant. Kubrick elected to bring the power of all four techniques of film style to bear on the problem of communicating the film's main ideas. Specifically, these are cinematography, mise-en-scene, editing, and sound. While all four are employed in significant ways to support the film's narrative, they also communicate in a very direct way the emotions, atmosphere, and symbolism that support the film's aforementioned three main ideas. To understand how this comes about, we must examine Kubrick's use of each technique in greater detail.

Cinematography: Kubrick's decision to shoot with 65 mm film yielded an image that is stunningly real on the screen. Even today, no other medium can come close to its level of resolution. Consequently, the audience was drawn into *2001: A Space Odyssey's* futuristic world far more easily and more readily accepted it as reality than it could with any previous science fiction film. With its wide 2.21 to 1 aspect ratio[23], combined with projection on a large screen, the 70 mm release print used in the Cinerama Super Panavision 70 format[24] offers an image that is an overwhelming spectacle. Framing faithfully places what Kubrick deems most important at center stage. Such shots often include tools, instruments, and displays and reveal an unprecedented level of detail and sophistication in their design. The angles from which the spacecraft are photographed repeatedly emphasize their grandeur and what a marvel of engineering they are. Close shots of the surfaces of spacecraft and landing facilities reveal intricacies of their structure that enhance the realism of these scenes to a degree far beyond that seen in previous science fiction films, as do the miniature process shots or movies occurring within the frame. These are shots of interiors seen through windows in spacecraft, the lunar landing facility at Clavius, and the shots of windows in the docking bay of the space station where we see the crew monitoring the approach of the *Orion* in Act II. Not only did this technique contribute to the realism of scenes involving spacecraft and landing facilities, it also gave the audience a dramatic sense of the scale of these structures and how they dwarf their human occupants.

During Acts III and IV, Kubrick uses both the subjective camera[25] and special wide-angle lenses to great effect. In Act III, the distorted images we see through Hal's eye are the result of shooting the scene from Hal's point of view with a fish-eye lens[26]. Frequent close-ups of Hal's electronic eye convey the impression that he is both an omnipresent as well as omniscient presence aboard the *Discovery*, but more importantly, that he is a character that needs to be watched more than trusted. In Act IV, both the point-of-view shots and distortion created by the lens help us empathize with Bowman's "stranger in a strange land" experience as he explores the alien suite. The long, wide-angle shot of the space pod after it arrives at the alien bedroom is an excellent example[27]. The surroundings look familiar, yet through the lens, space appears distorted; lines that should be parallel are curved in this place. We sense, as Bowman does, that things in the alien world are not normal or real in the sense that we understand reality.

Most importantly, Kubrick mastered the art of the moving camera in a way that gave us scenes which were more than merely dynamic. Kubrick, working with cinematographer Geoffrey Unsworth and the innovative camera motion apparatus developed by Wally Veevers, presents scenes in which the motion relative to the subject is so well integrated with the music that the aesthetic effect of both is greatly amplified.

Mise-en-scene: Regarding the various sets and properties which include spacecraft exteriors and interiors, the moon base, the landing facilities, and the lunar landscapes all obtained a level of realism never reached before and rarely matched in any science fiction film since the advent of *2001: A Space Odyssey*. The same can be said for the shots involving crew personnel moving and working in zero gravity. The impact that this technique had on audiences was not lost on directors of future science fiction films. After *2001: A Space Odyssey*, exterior and interior sets as well as properties became far more detailed and realistic in science fiction films, particularly those that involved space travel.

Mood lighting and the use of tones and colors in general emphasize the collective thinking processes during the evolution of the human race. Earth tones and blue skies dominate in the time of the primitive ape-men as instinct begins to evolve into intuition. By the twentieth century, ape-man has evolved into computer-man living in a world of divalent logic with zeros and ones reflected in the absolutes of black and white in outer space. To move beyond this, Bowman must pass through a star gate/threshold of dazzling contrasting colors where he encounters the world of the super-intuitive aliens. In this realm, intellect has evolved beyond machine logic as the aliens seem to control their environment by thought alone.

Kubrick may have also used color to suggest a link between Hal's electronic eye and the danger that arises from his irrational fears. The red seen in the lens of Hal's eye is rarely used on the sets in *2001* with one prominent exception, the interior of the Main Logic Center, which we see only at the end of Act III when Bowman proceeds to terminate Hal.

The behavior and dialogue of the human characters remains vapid throughout the middle sections of the film. By Act III, the behavior of Poole and Bowman stands in stark contrast to the behavior of Hal, making them appear dull and uninteresting as individuals. Poole's zombie-like appearance and behavior during the scene where he receives the birthday video from his parents provides a striking example. This strategy emphasizes the reciprocal relation between the evolution of the machines and the evolution of the human race: as one becomes more machine-like the other appears more human. By directing his actors to behave this way, Kubrick underscores their characters' need to rediscover their intuitive ability and revive their humanity. We also see that by not letting us come to know the characters in depth, Kubrick brings out their mythological dimension wherein the characters in their various roles and times need to be seen as representatives of an evolving race in a drama that is far more about the race than it is about any individual.

Tools and technology play a strong supporting role in the narrative, stealing many of the film's scenes. This allows us to see them as characters more than we normally would. Hal, in particular, is far more than an advanced computer assisting with operations on board the Discovery. He is a talking, thinking, feeling machine who is in every way a principal character in the story, as well as its most sinister villain. Some writers have gone so far as to claim that, with the exception of Floyd's daughter, Hal is the most human character in the film. However, I don't find this an accurate description of what we see during the plot, nor is it in keeping with Kubrick's futuristic myth. Hal appears to express more emotion than Dave or Frank, but the emotion in his responses sounds more programmed than human. It is as if we are hearing what someone thinks would be the most appropriate emotional response to the dialogue and behavior of his human crewmates. Not until his conversation with Dave Bowman about the "extremely odd things" that happened during preparation for the mission, does Hal's dialogue begin to sound spontaneous and genuinely human. Hal also proves to be emotionally unstable, but animals exhibit emotions and emotional instability also, so this is not a uniquely human trait. Dave Bowman, on the other hand, tends to be reserved and doesn't exhibit much emotion until he is forced to deal with Hal's breakdown and the murder of his crewmate, Frank Poole. Yet, Dave was the artist in the crew, not Hal; and we need to ask, where were Hal's humanity and feelings for Mr. and Mrs. Poole when he methodically and cold-bloodedly murdered their son and the three hibernating astronauts who were no threat to him? To his end, Hal never shows any genuine remorse, only fear of being terminally shut down.

Certainly, Hal is a character with a crucial role, and his cunning and highly developed artificial intelligence makes him a formidable antagonist to Bowman. However, without the benefit of repeated viewing, many might miss the way in which the subtleness and restraint of Keir Dullea's portrayal of Bowman allows us to see that beneath the cool, professional surface of this character lies a complex, courageous, and truly human personality. While Hal calculates, analyzes, and reacts, Bowman reflects, creates, and foresees.

With this pair of characters, Kubrick has established a fascinating counterpoint relationship where one will devolve and mentally disintegrate while the other will evolve and supersede his Shadow by drawing on his intuitive side and using it to regain his humanity. In Dave Bowman, Kubrick has fashioned a character who has the qualities necessary to not only triumph over the machine, but to become more whole as a human being and move on to the future.

Editing: Cuts that span huge gaps in time in *2001: A Space Odyssey* might seem to make the film overly episodic or even fragmented. However, in a story spanning millions of years, splicing together large sequences spaced widely apart in story time helps us to understand that the ape-men, Heywood Floyd, Dave Bowman, and the Star-Child make one super-character with a role that spans the entire narrative and encompasses the history of the human race. In the transition from the first act to the second act, Kubrick uses a stunning graphic match cut from the bone-tool to the man-made satellite in order to link the two in a way that underscores the astounding transformation that evolution has brought about as we witness it all from a God's-eye view. A simple cut also serves as the transition from Dave's awakening from a state of shock at the end of the star gate sequence to his disorienting arrival at the alien suite. Using the cut as a transition impels us to form associations between the images and scenes more than dissolves or other elliptical transitions would. Near the end of the film, jump cuts defy cause and effect and undermine conventional continuity, helping us to empathize with Bowman's fear and bewilderment in an alien world. The jump cuts also add considerable ambiguity to the question of how much time Dave has actually spent in the alien suite. His stay could have been for a few minutes or many Earth years.

Shot duration and timing of cuts with subject movement are two of the most important elements that determine the pacing of any film. In *2001: A Space Odyssey*, many shots such as those of spacecraft exteriors and interiors are held for a much longer time than would be necessary for showing narrative action or establishing setting. However, this additional time allows us to enjoy the formal qualities of the shot and contemplate the progress of the subject's evolution. The gradual transitions from black allow for reflective associations between sequences where the human race has been largely marking time rather than making dramatic leaps, as in the transition from the rising sun in the opening title to the rising sun at dawn in the desert and from the prehistoric monolith unearthed on the moon to the skeleton-like *Discovery* drifting towards Jupiter. Duration and timing of transitions are essential for both the abstract sequences and the dance sequences (e.g., the beginning of Act II). Image form and movement from cut to cut must flow in rhythm to the music or the scene would immediately lose its emotional and aesthetic impact.

Sound: Kubrick's emphasis on the sound of Bowman's steady breathing as he reenters the *Discovery* at the end of Act III underscores the character's effort and determination to survive Hal's efforts to end his life, but on an allegorical level suggests a resuscitation of the human spirit that has begun to take place. However, no other stylistic device in the film contributes more to the tone of the images on the screen than the music Kubrick selected for the soundtrack[28]. To recapitulate, *The Blue Danube* celebrates the precision, grace, and triumph of man's technology. Ligeti's music, particularly *Requiem for Soprano, Mezzo Soprano, Two Mixed Choirs & Orchestra,* and *Lux Aeterna,* evokes a sense of mystery, eeriness, and awe the characters feel as they encounter the work of super-intelligent, extraterrestrial life. By the beginning of Act IV, we realize how both the radiant tone clusters of *Requiem* and the luminous, angelic voices of *Lux Aeterna* have become themes that have introduced the deep, cosmic mysteries in *2001: A Space Odyssey.* Yet, as *Lux Aeterna's* title would suggest, when the human race addresses these mysteries the protagonists move into light. In fact, the protagonists and we as members of the audience are constantly moving into light throughout Kubrick's space odyssey, never into darkness. A third Ligeti composition, *Atmospheres,* accompanies Bowman's dazzling intergalactic ride to the alien world. Here Ligeti's brilliant atonal orchestration, structured primarily on the basis of instrumental timbre, communicates the strange and marvelous quality of Bowman's journey and integrates perfectly with the flow of Kubrick's abstract patterns of light and color.

The brooding, sorrowful "Adagio" from Khachaturian's *Gayane Suite* heard at the beginning of Act III suggests that something isn't right on board the *Discovery* and foreshadows that something dark and tragic may lie ahead for the crew. Indeed, if we look back on the events that unfolded in Part 1 of Act III as a whole, the "Adagio" becomes a requiem for the death of the human spirit. This idea is brought out poignantly with a twist of Kubrickian irony in the scene where Frank Poole watches his parents struggle to celebrate his birthday back on Earth. Khachaturian's "Adagio" (non-diegetic music[29]) is overlaid with Frank's parents singing "Happy Birthday" (diegetic music). The effect this combination of the two pieces of music creates is sad and disturbing. Add to this the irony of Hal's artificially sweet voice and you have a milieu that completely underscores Frank's dehumanizing detachment. Music and sound effects are used to create irony again when the perfect, more human-than-human machine, after breaking down and committing murder, offers us a love song as it watches itself being terminated. Written in three-quarter time, "Daisy Bell" [30] recalls *The Blue Danube* that opened Act II, but this is no celebration for the machine. "Daisy Bell's" whimsical lyrics, sung by Hal as his brain is disconnected and his voice becomes slower and much lower in pitch, reduce this once intellectual titan to something that sounds like a child's broken toy.

Finally, the opening chords of *Also Sprach Zarathustra* that open and close the film evoke a sense of majesty, awe, and optimism at those seminal moments when humans celebrate the triumph of their inventiveness, stand before a threshold to a new adventure, or contemplate their destiny. These examples demonstrate that Kubrick has used music, particularly non-diegetic music, in a way that goes far beyond the common convention of providing some background atmosphere or supporting the emotional side of the narrative. Music informs *2001: A Space Odyssey* and, through motifs, systematically cues us to understand better its symbolism and ultimately its grand ideas. Music paces the scenes and the transitions between scenes throughout the film, from the opening title to the full shot of the Star-Child at the end of the film. Music and the images choreographed to it work on us directly, whereas words from characters and their facial expressions simply wouldn't be enough. Kubrick understood that even with the best actors these would fail, becoming a kind of emotional noise that would distract from, rather than add to, the power of the scenes.

Scenes where techniques of style are integrated to work in concert to evoke a response or communicate an idea are the ones that have the greatest impact. This is particularly true of the many motifs that occur during the film, such as the alignment of three celestial objects or two celestial objects and the monolith. Here cinematography, mise-en-scene, and sound work together to create a profound impact on the viewer. At the most general level, this pattern is communicating the idea that specific, rare celestial alignments signify critical or special moments in time which possess a quality or meaning in synchronization with the quality and meaning of what the human race is experiencing at that time. Mise-en-scene gives us the alignment and triad, and cinematography enhances the significance of the pattern using camera angle, image size, and high-resolution film stock. The music in the soundtrack reinforces the triadic structure while capturing the tone and emotion of the moment.

A spectacular example of Kubrick's use of all four techniques of style working in concert may be seen in the series of shots that open Act II. Make no mistake about it, this is a dance sequence and every move was painstakingly choreographed to the music (sound) primarily through very careful selection of footage and editing. Camera movement and subject movement are exquisitely timed so that we feel the joy of a celebration as spacecraft waltz to *The Blue Danube*. Camera angle, image size (cinematography), lighting, color, blocking, and set design (mise-en-scene) enhance the communication of this feeling even more. Selection, duration, and timing of the cuts (editing) from the satellites to the earth, moon, space station, and the *Orion* allow us to take in all the various participants in the dance while maintaining the proper overall pace the sequence requires.

Even without the benefit of music, dialogue, or voice-over narration, Kubrick was often able to create powerful, affecting scenes by engaging the talents of sound designer Winston Ryder. Hal's murder of the hibernating astronauts provides an excellent example of such a sequence. Before Dave Bowman returns to the *Discovery* with Frank Poole's body, Kubrick needed to show that Hal's villainy could extend far beyond murdering Frank Poole and make us fearful about Dave's vulnerability while he is out in the space pod, well beyond the *Discovery*, attempting to retrieve Frank's body.

The sequence begins with a cut from a close-up of Dave piloting the space pod to a close-up of Hal's red eye, which has taken on a menacing quality since the fatal event where Frank was struck with the out-of-control space pod. The next shot shows two empty chairs from Hal's point of view in the centrifuge room reminding us that Frank and Dave are absent and that Hal is now free to do as he pleases. Kubrick cuts from a hibernating astronaut to the display panel monitoring the astronaut's six life support functions. One by one, the monitor for each of the three astronauts displays their life functions gradually diminishing to flat lines. To our horror, Hal has taken advantage of the helpless state of the hibernating astronauts, asphyxiating them as they lie sleeping in their modules. We witness the progress of Hal's lethal plot as it proceeds through three stages indicated by the messages displayed on a flashing red panel: Computer Malfunction, followed by Life Functions Critical, and finally Life Functions Terminated–alarm, frantic alarm, hopelessness. The last three shots are a long shot of the three astronauts, each lying dead in his hibernation module, a close-up of Hal's red eye, and a wide shot of the pod bay with only one space suit hanging on the wall (reminding us of Bowman, now the sole surviving crew member). Using clever cutting, affecting shots, and colorful graphics in the displays, combined with Ryder's intriguing sound effects, Kubrick has created a sense of mounting suspense, alarm, helplessness, and a lasting image of Hal as a deliberate, cold-blooded killer.

While this is not an exhaustive treatment of Kubrick's style, I have tried to offer some detail on the more salient elements and techniques employed in *2001: A Space Odyssey* with the hope that this will be sufficient to persuade the reader that it is *2001*'s style which gives unity to the film's diversity of forms. Style carries the themes of *2001.* Through its 65 mm cinematography, set design, editing, music, patterns of color, the motion of objects synchronized to music, and most importantly, the meaning the objects in motion share with the music, *2001* consistently conveys the idea that the future is really happening; that man's tools will evolve to a state of majestic beauty and attain artificial intelligence; and that our evolution, the evolution of our technology, the exploration of space, and the possibility of encountering alien life are awesome, soul-jolting experiences. Amidst all of its extraordinary detail and visual information, *2001*'s style with its motifs, archetypes, and super motifs keeps us focused on the broad strokes so that we never lose sight of the forest for the trees. Kubrick has reminded us of the evolution of cinema itself, which began as a way of telling stories using images and music (whether played live or from recorded media). Kubrick's style demonstrates that these techniques remain viable and effective to the present day.

Unanswered Questions and Mysteries

We have seen how patterns, archetypes, and style not only inform *2001: A Space Odyssey*, but communicate essential information related to both its story and Kubrick's three main ideas. Kubrick does not explain everything, however, and a number of unanswered questions and tantalizing mysteries remain.

The function of the monolith has probably been the most debated mystery in the film. Most likely, it was placed on Earth and the moon to serve as a monitoring device because the aliens would have needed a means to keep track of our progress until we evolved to the point where they felt it necessary to bring a representative of our race to their world for closer study. Yet, many writers and viewers believe it played a much more active role. They maintain that the monolith was a device that the aliens used to alter human thinking[31], if not human evolution, pointing out that it is present at three very critical times during the plot. I believe the best way to ascertain the function of any element in a film is to first study its use in the film itself.[32] We can rule out coincidence because a correlation between the appearance of the monolith and key events in the history of human evolution definitely exists and Kubrick has placed it very prominently and dramatically within his film. There is no evidence of any mechanism bringing about changes in either the apes or the humans or even communicating messages to them, but this cannot be ruled out. However, Kubrick may have been suggesting that the relationship between human evolution and certain rare celestial alignments was synchronous and that the aliens had knowledge of such phenomena. Synchronous relationships between events are those that are acausal, in certain situations repeatable, and possess some quality or meaning in common. [33] [34]

This commonality may also extend to periods of time which in turn may be signified by the periodicity of specific celestial alignments. Suggesting that synchronous relationships could explain critical points or shifts during the course of human evolution would have been an extraordinary insight on Kubrick's part. He also implies, through the behavior of his characters, that twentieth-century science was largely unaware of these ideas. Steeped in the deterministic, cause-and-effect scientific method of the twentieth century, people like Heywood Floyd, Dave Bowman, or even Hal would have regarded this kind of thinking as a radical departure. In this sense, Kubrick has anticipated a new paradigm for twenty-first century science[35] [36] – one that could embrace the significance of time-related, acausal phenomena.

Beyond the monolith's role as an element in a synchronistic event occurring at a critical point in the history of human evolution, it appears that Kubrick also felt it was important to emphasize that both the monolith and the apes' encounter with it convey the feeling of a numinous, aesthetic experience that would inspire them in a profound way. This is a point that has been overlooked by many writers and critics who have placed, perhaps, too much emphasis on the apes using their bone-tools as weapons to kill tapirs for meat or to kill other apes to gain control over resources such as the water hole.

The idea of a numinous experience is powerfully evoked by Ligeti's *Requiem* which is heard in the soundtrack throughout the scene in which the monolith first appears. Initially, the apes awaken and respond to the monolith with a mixture of curiosity and fear. As they gather around it, they become extremely agitated, yet are drawn to it irresistibly. They touch the monolith and cling to it, admiring its smooth surfaces and precise right angles. Something has begun to stir in their primitive minds that allows them to appreciate the beauty of its geometry, its symmetry, and the precision involved in its fabrication. From this time forward, these qualities will inform their work for millennia to come. We have witnessed a response unlike anything we would see from the leopard, the tapirs, or any other animals if they were to respond to the monolith at all. It is a truly human response and it is far more advanced than the act of picking up a bone and using it as a weapon to hunt or kill. To extend this numinous experience in time, ape-man must create and continue creating, evolving into twenty-first century human and beyond.

A second mystery is the mission or purpose of the Star-Child. Everything points to the likelihood that he represents, at the very least, a new step in the process of human evolution which, in Kubrick's view, seems closely allied to the holistic idea of evolution being a process unfolding in "positive" or "creative" time, as Bergson characterized it. The following quote from an article on holism, written by philosopher Jan C. Smuts, is enlightening:

> Wholeness or holism characterizes the entire process of evolution in an ever-increasing measure. And the process is continuous in the sense that older types of wholes or patterns are not discarded, but become the starting points and the elements of the newer, more advanced patterns. ...The whole is creative; wherever parts conspire to form a whole, there something arises which is more than the parts.[37]

I think it is safe to assume the Star-Child has the potential to greatly alter things on Earth, including the future of mankind. Given Kubrick's philosophical leanings, we might speculate that the Star-Child will serve as a liaison between humans and the aliens. Perhaps he will show humans how to build even greater machines, or save humans from being dominated by their machines. Some might argue that a darker end is at hand. Yet, while destruction, decay, and death remain a part of the human odyssey, which Kubrick has not failed to show us, neither nihilism nor hopelessness about the future of mankind is a theme in *2001.* Indeed, we have witnessed a resurrection at the end of the film, a resurrection/rebirth indicating the most likely mission for the Star-Child will be the re-humanization of the human race. Campbell's research in comparative mythology also supports this. In his description of the mythological hero's return Campbell writes:

> "the hero has died as a modern man, but as eternal man–perfected, unspecified, universal man–he has been reborn. His second solemn task and deed therefore... is to return then to us, transfigured, and teach the lesson he has learned of life renewed." [38]

While any of the aforementioned alternatives could serve as a possible epilogue, open endings remain unsettling for some viewers. However, a story about evolution is necessarily about a continuing process that may have no end. Thus, Kubrick's ending serves his story better than the alternatives; it remains true to its theme–the ascent of the human race–and maintains our curiosity.

A third mystery often discussed is the cause of Hal's breakdown. The first indication we have that something is troubling Hal appears when he begins to question Dave about whether Dave "might be having some second thoughts about the mission." As the conversation proceeds, Hal embarrasses himself in his attempt to find out what Dave might know about the true purpose of the mission, but then he senses some form of information that leads him to conclude that the AE-35 unit will fail. This of course takes the conversation in a new direction, but it becomes one which allows Hal to feel he is on firmer ground with Dave and that Dave's confidence in him has not diminished. That is, until Dave's own analysis of the circuit and further investigation into the matter by the HAL 9000 on Earth both fail to confirm Hal's prediction. At this point we must ask, has Hal made an honest mistake in predicting the failure or has he invented this problem to test the trust and loyalty of Dave and Frank?

Either way, once Hal discovers that Dave and Frank no longer trust him, he decides that he must act to prevent the humans from possibly shutting him down. For Hal, things are either true or false, they either fail or don't fail, and humans may either believe what you tell them and remain loyal to you or not. His breakdown, therefore, appears to be the outcome of following Heywood Floyd's instructions that he always be truthful, but not truthful with regard to the purpose of the mission–a conundrum for Hal that he can neither solve nor live with as the mission proceeds. For all his sophistication as an A.I. entity, Hal's logic remains largely divalent. This seems to render him incapable of maintaining what little humanity he might have developed which, in turn, could have allowed him to find a more reasonable, less deadly solution to the problem of Dave and Frank not trusting him.

A number of writers have commented on the film's raising questions about free will and whether humans or the aliens are really in control of the destiny of the human race.[39] While events in the fourth act strongly imply that the aliens acted, at least, as midwife in the birth of the Star-Child, this is not the same thing as determining the destiny of the human race nor does it preclude free will on the part of this new species. There is simply no evidence that either the ape-men or the Star-Child were mind-altered to behave or think according to someone else's plan with no free will of their own or the ability to conceive an original idea. Again, judging by the film itself, Kubrick has avoided giving us a firm answer, but screenplay co-writer Arthur C. Clarke offered insight into Kubrick's intentions in a general way when he stated, "If you understood *2001* completely, we failed. We wanted to raise far more questions than we answered." Kubrick wanted his audience to think about these questions and ideas and come up with some interesting ideas of their own, and though he was often ambiguous in *2001*, he was not equivocal. In the aforementioned instances where *2001: A Space Odyssey* does not offer us solid answers through dialogue or text, the patterns, motifs, and various archetypal forms continually express an attitude that is positive and hopeful. In the end, unresolved mysteries notwithstanding, *2001*'s images and music leave us filled with a sense of wonder.

The difficulty some viewers and critics had in understanding and/or appreciating *2001: A Space Odyssey* that I referred to at the beginning of this essay most likely stems from Kubrick's deviations from the conventions of the genre and also what is known as the "classical Hollywood mode" of narration. [40] The film's use of multiple formal strategies and the allegorical quality of the story are two examples. While neither the use of multiple formal strategies nor allegorical narration was new to film by the late sixties, Kubrick's innovative combination of these methods was more than American audiences were prepared for. It was something they might have expected to see in foreign art cinema, but not in an American science fiction film.

A third important cause of audience dissatisfaction was Kubrick's avoidance of several conventions and norms of expression in *2001: A Space Odyssey*, particularly expression through character behavior. By the early sixties, American-made science fiction films had developed a set of conventions and norms that became popular with audiences and were generally respected by screenwriters and directors of science fiction films. The protagonist or hero in the film was usually played by one character; there was often a subplot involving a romantic relationship between the leading male and female characters; aliens were viewed as a threat to humans and the earth; exposition normally came from character dialogue; and the style of spacecraft, properties, and costumes took precedence over making them look realistic. Finally, if anything mysterious, incomprehensible, or spectacular occurred, audiences expected to see the main characters talking about it, expressing their feelings, and sharing their thoughts. Yet, Kubrick reduced these forms of expression to a minimum and avoided all of the conventions mentioned above. Consequently, many left the theater feeling unsatisfied and that the film was lacking something.

Most directors would have followed the traditional path of communicating ideas and emotions through actors acting out their character's reaction to an extraordinary event or experience. However, by minimizing traditional character behavior and dialogue, by eliminating voice-over narration, and presenting more shots from the character's point of view; Kubrick's style allowed striking visuals, archetypal symbols, and music to shine through, hitting us directly and intensely where our aesthetic sense grasps the essence of his ideas. Had he relied on actor's expressions and emotional responses, much of the impact would have been lost, owing to the indirect interpretation of the event we get through the actor portraying the character's response. This is similar to listening to a person describing Michelangelo's statue of David or a performance of Stravinsky's *Le Sacre du Printemps*. No matter how descriptive their language or how much emotion they are able to muster, the experience would not be the same as it would if you were viewing the statue or listening to the performance in person. The aesthetic impact is missing.

Even today, few films from any genre can be compared to *2001: A Space Odyssey* in this regard. Kubrick's mode of narration demands a greater awareness of symbols, archetypes, and style because it is through these that he reveals the film's themes. *2001* also requires a greater sensitivity to music and the non-representational or abstract visual arts because the film's aesthetics work on these levels as well as on traditional narrative film levels. While Kubrick doesn't offer a "how" or "why" for some of the more significant events in the film, neither does he give us the studied ambiguity often observed in foreign art cinema. If the answers are not available in conventional forms of text such as expositional character dialogue or voice-over narration, the film's coherent archetypal symbolism combined with an abundance of associational links keeps our minds engaged and provides us with clues. Rather than merely thwarting our expectations, whether they derive from our experience with the science fiction genre or classical Hollywood mode of narration, Kubrick's style constantly reinforces the symbolism and meaning of his images, presenting us with a variety of opportunities to form associations, interpret them, and draw our own conclusions. Through his consistency, Kubrick encourages us to discover a new path to meaning in narrative film. Although he took considerable risk by putting his ideas and what he called "the feel" of his film ahead of character development and story, Kubrick's principal intentions were never truly obscure and are perhaps best summed up in the following quote: "What I'm after is a majestic visual experience."[41]

Conclusion – Kubrick's Gift

For my final topic, I will take a closer look at what I believe is Kubrick's most original and important innovation not only to the science fiction genre, but to narrative film in general. This is a stylistic technique that I have referred to earlier as music-driven imagery in *2001: A Space Odyssey.* We need to replace this rather vague phrase with one that is more precise and scientific. My term for this kind of phenomena is musicography or the musicographic image.[42] A musicographic image is one in which the formal or abstract qualities of the image, such as subject size, color, velocity, shape of path, speed of rotation, darkness or lightness, are integrated with the music so completely that the aesthetic experience of the audience is amplified non-linearly. It is also important to understand that musicography is not the same as the phenomenon known as synesthesia or chromesthesia. This has been investigated and defined as response or visualization that occurs within the subject's mind when listening to music.[43] Musicographic images are created outside of the mind and perceived through our senses of hearing and sight. Although a thorough discussion of this fascinating topic would be far beyond the scope of this essay, I hope to give the reader a basic idea of the concept and its importance to film and many of the other visual arts.

Given its intense aesthetic impact, the musicographic image is easy to recognize, but difficult to describe in words. As a phenomenon better understood through demonstration rather than explanation, I would encourage the reader to review and study three prime examples and a fourth very brief, but nonetheless striking example that occurs near the very end of the film. These are the opening title, the *Orion's* voyage to the space station that opens Act II, the star gate sequence in Act IV, and the full shot of the Star-Child that occurs at the end of Act IV. Hopefully, as it becomes better understood, the use of musicographic imagery will expand in film and other visual media in the future.

The opening title of *2001: A Space Odyssey* clearly demonstrates that timing and rhythm of movement of the visual components are extremely critical in the presentation of a musicographic image. The title begins with an extremely deep C played on an organ, the first note of *Also Sprach Zarathustra.* Observe how this is integrated with the dimness, visual mass, and velocity of the celestial bodies as they begin to rise. Accompanying their ascent are the first three notes of the fanfare, C G C^{8va}. This is punctuated by a C minor chord, just as the brilliant light of the sun breaks over the earth's horizon. The visual image reinforces the great tension created by the sudden major-to-minor chord change at the end of the first fanfare. The second fanfare ends on a brighter, C major chord, as the sun continues to rise, shedding more light on the scene. At the beginning of the third fanfare, the words "Metro-Goldwyn-Mayer presents" burst on the screen in white letters, balancing and establishing a base of brightness towards the bottom of a form that has evolved into a pyramid. The sun has now risen three quarters of its diameter above Earth. As the fanfare comes to a climax with a triumphant F major chord and a cymbal crash, the second, larger credit, "A Stanley

Kubrick Production," flashes across the screen. The full orchestra now plays the final part of the theme returning to the original key of C major, climaxing with a greater cymbal crash, as the title "2001: A Space Odyssey" explodes out to the full width of the screen in still-larger white letters with the full sun crowning the top of the triangular form. Having ascended to the light, the final C major organ chord reverberates and fades as the image fades to black for the transition to the first act.

The opening sequence of Act II presents some of the film's most exquisite choreography, but to see this sequence musicographically, we must look at its shots more as abstract compositions than representational cinematography. Observe the abstract qualities of the spacecraft in these scenes, specifically their shape, velocity, rate of rotation, and light color against the black background of space. The sequence begins with a shot of a satellite in orbit around Earth as Strauss' *The Blue Danube* waltz begins slowly and quietly. At first, everything seems to be drifting, including Floyd's pen. Gradually, as the waltz rhythm of the music is asserted, the various subjects of the shots take up the dance as their speed and rotation are synchronized with the rhythm. In subsequent shots, both the speed of the camera pull-backs and the *Orion* gliding across the screen or deeper into the background are closely integrated with the gliding movement evoked by the waltz. In the second triad formed by the space station, the moon, and the *Orion*, notice the speed of rotation of the wheel-shaped space station and *Orion* gliding to it–both in synchronization with the three-quarter meter of *The Blue Danube.* As the music swells, Kubrick cuts to a side view of both spacecraft and these two shapes loom much larger on the screen. In the next shot, the waltz reaches its crescendo as the camera tracks in even closer, moving within the outer rim of the rotating station. The dramatic climax of the expanding, white, rotating mass and the music is perfectly integrated.

The star gate sequence begins as Bowman's space pod approaches the complete blackness of the rectangular threshold, with Ligeti's *Requiem* rising in the sound track. After he enters the abyss, the first streams of colored light pass by, creating a sense of rapid acceleration through a great distance. Patterns of brilliant, streaming colors shoot out at us intercut with shots of Bowman's face shaking with patterns of color from the illuminated instrument console reflected off his visor. The shaking increases to a vibration that becomes so rapid that Bowman's features eventually blur and merge with the vibrating patterns of color on his visor. The notes and dissonant harmonies performed in Ligeti's music resonate and stream out at the viewer, as do the vibrant waves and shards of contrasting colors. As the music swells and the multiple, often dissonant tone clusters sung by the chorus clash and shatter, the patterns of light also swell and shatter into multiple colors. *Requiem* then transitions to *Atmospheres* and we flow through planes and streams of fluid color on to the beginning of creation and the birth of galaxies as they expand and flow in time. Overall, we experience a metaphor in color, form, and music for Bowman's and our astounding journey through space and time to the world of the aliens.

At the end of Act IV, the *Zarathustra* fanfare used during the opening title returns as the Star-Child passes through the monolith/star gate to return to Earth. The frame descends from the moon to a side shot of the Star-Child in orbit around the earth, then cuts to a full-frontal shot synchronized with the same triumphant C major theme and cymbal crash that was used in the title when the words "2001: A Space Odyssey" burst on the screen. The final C major organ chord reverberates and fades with the image of the Star-Child–enigmatic and majestic.

For a small but significant fraction of the audience, the intensity of this kind of aesthetic experience exceeds the impact of anything that could be felt as the result of either the images or the music acting alone. It is direct, stunning, and mesmerizing. Yet, exposure to this kind of imagery was still a very new experience for most members of the film-going public. Musicographic images had not been used in any science fiction film prior to *2001: A Space Odyssey,* although they had been attempted in some experimental work and in the abstract films created by filmmakers and animators such as Oscar Fischinger, John and James Whitney, and Mary Ellen Bute as well as some of the animation and dance films directed by Norman McLaren. Certainly John Whitney's work in *To the Moon and Beyond* [44] (1964) caught Kubrick's attention, and it is also quite likely that Kubrick had seen and been impressed by the mesmerizing opening title to Alfred Hitchcock's *Vertigo* (1958), which was designed by Saul Bass and John Whitney, who composed their visual concepts to Bernard Herman's brilliant score.

Among films distributed for general theatrical release, the most notable with regard to attempts at creating musicographic images was Walt Disney's animation feature, *Fantasia* (1939). Considering the thought and intention behind Kubrick's work, we can see how the visual images in *2001* derive more from *Fantasia* than from any earlier science fiction film or drama. *Fantasia* also mixed narrative with abstract form during its plot; however, its plot was deliberately divided into a number of distinct animated episodes. Its unity of form derives from Disney's original conception that each episode drawn by the film's animators would be inspired by the piece of classical music selected for its soundtrack. Comparing the films with each other as a whole, however, many of the scenes in *2001* were also composed to the music used in the sound track and both are about and succeed because of the successful orchestration of visual images and music together. Unfortunately, neither the talent of Disney's team of animators, which included Oscar Fischinger, nor Kubrick's talent and success with this technique, has yet to be grasped by the majority of film critics and film historians. Consequently, both films remain unique, isolated, and largely underappreciated for their special artistic contribution to films made for general theatrical release.

Kubrick did receive letters from members of the audience who were deeply moved by his musicographic images. A response that was particularly creative and perhaps the one that had the greatest impact on Kubrick was written by Stephen Grosscup of Santa Monica, California shortly after the film's release. The following excerpt represents a fraction of his original letter, but it conveys Mr. Grosscup's main insight and a description of his creative project that Kubrick's work in *2001: A Space Odyssey* inspired.

Quoting from Stephen Grosscup's letter:

> It's a film of incredible irrevocable splendor. ... What if, I thought, someone wanted to see Beethoven's Ninth Symphony? Would you show them an orchestra and a chorus? Or what? Then I thought, what if someone filmed music? What if I filmed music?... I went out and filmed the third movement of Gustav Mahler's Seventh Symphony. It worked. It worked beautifully. ...it is doubtful that ever in my own lifetime will I be able to shoot a natural scene involving three planets. I knew instantly that you had too much reverence for Richard Strauss to tamper with the score– which meant that you had to put the film to the music–and you did–three times!... I have since done a great deal of thinking–and preparing–about 'filmed' music. I have gone so far as to envision a day when there will exist a new kind of artist who is both composer

> and photographer and who brings about a new form of art. But that day I think is a long way off, but maybe by the year 2001...[45]

Stanley Kubrick's reply:

> "Your letter of May 4th was overwhelming. What can I say in reply?"[46]

Indeed, Stephen Grosscup really got it. His insight is as original and remarkable as it is far-sighted. He has paid Kubrick the highest compliment a filmmaker could ever be paid. I leave you with the thought that we are not at the end of this journey; we are instead standing before a new threshold.

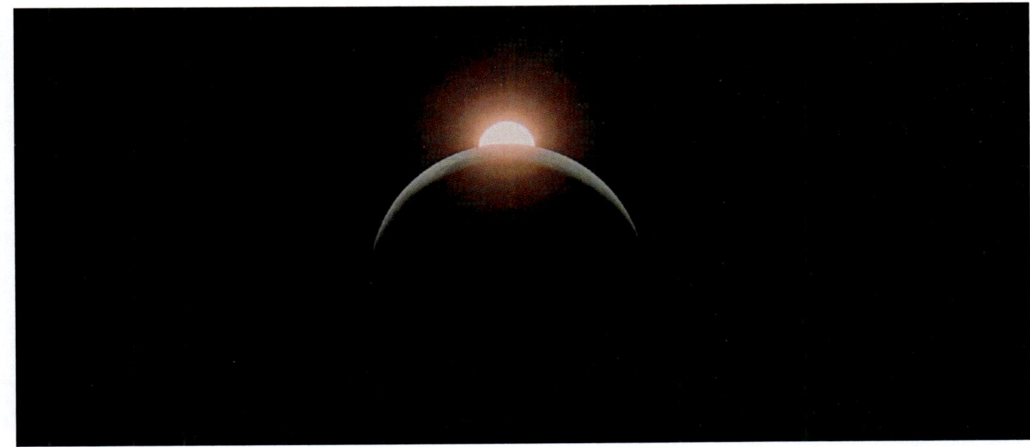

1. The earth and sun rise above the moon to form the first triad.

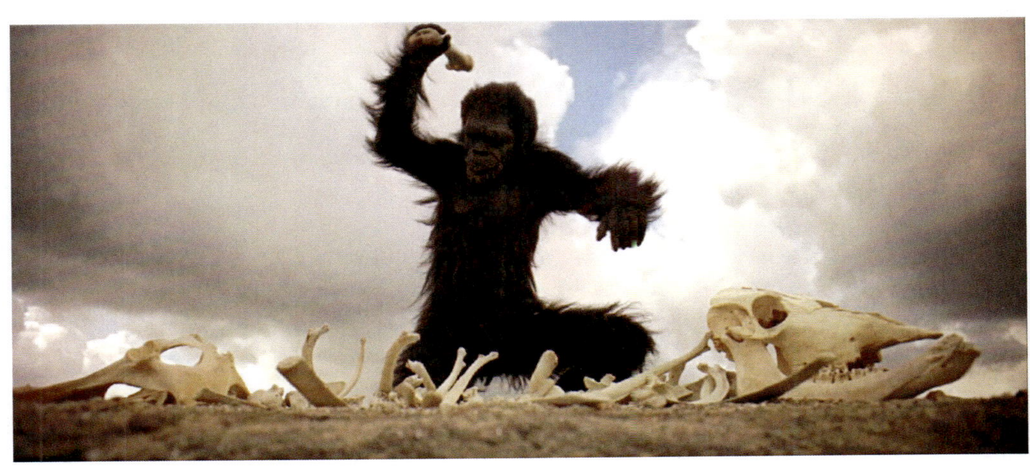

2. Moonwatcher discovers a bone may be used as a tool.

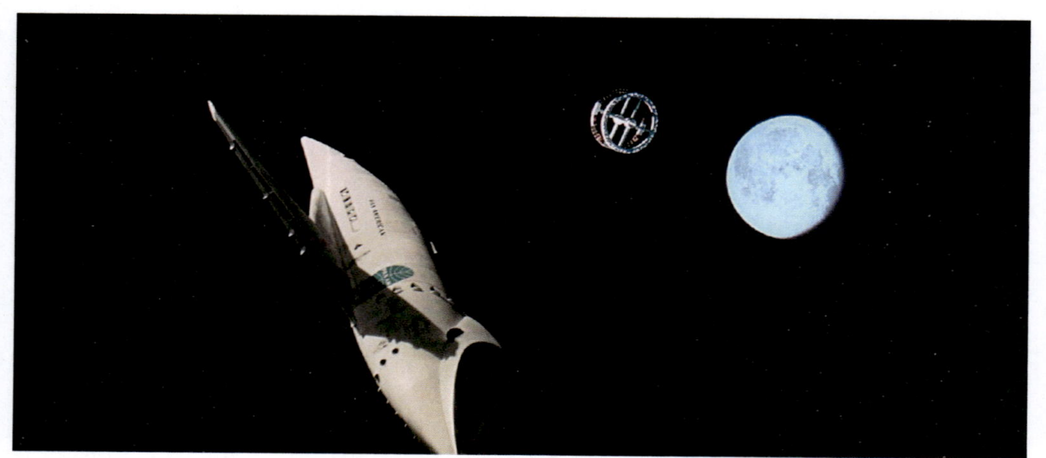

3. The *Orion* completes a triad as it glides to the space station.

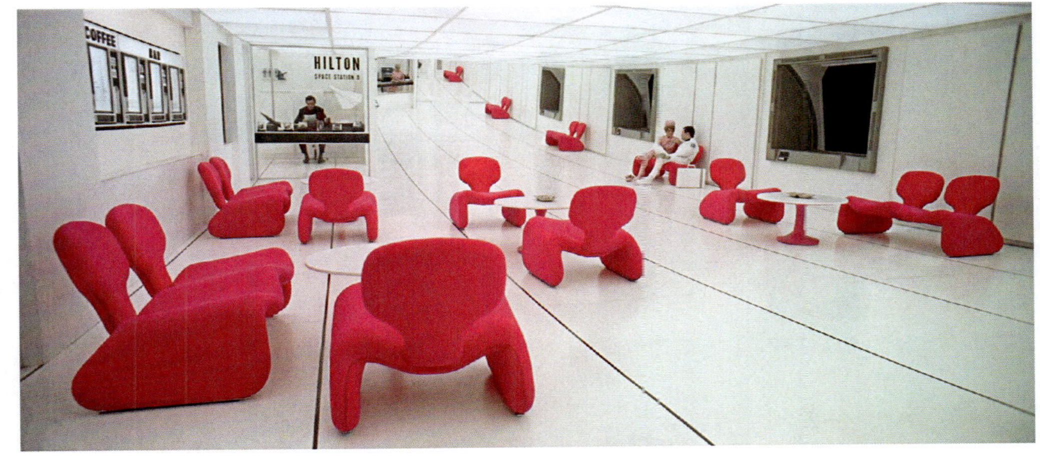

4. The stark interior of the space station's hotel obby.

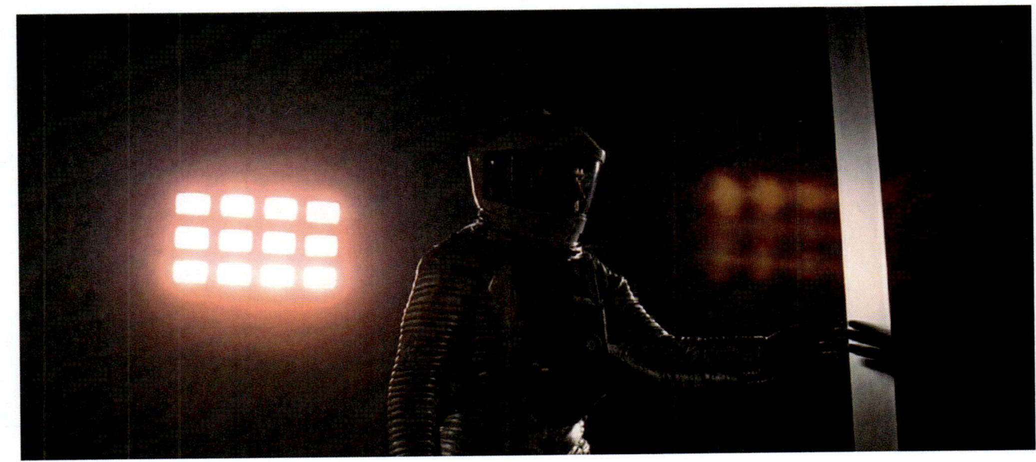

5. Heywood Floyd touches an alien artifact and his past.

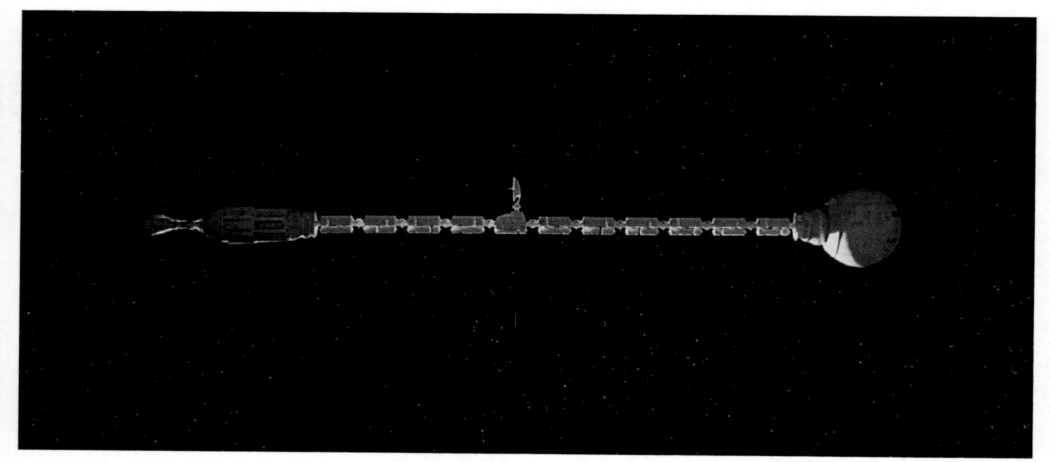

6. The skeleton-like *Discovery* drifts towards Jupiter

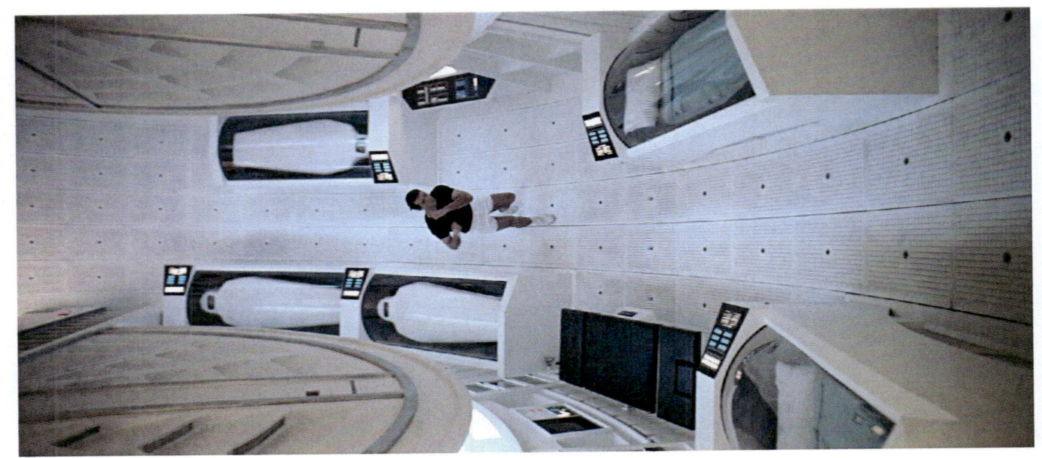

7. Frank Poole passes his sleeping comrades as he runs in circles boxing with an invisible opponent.

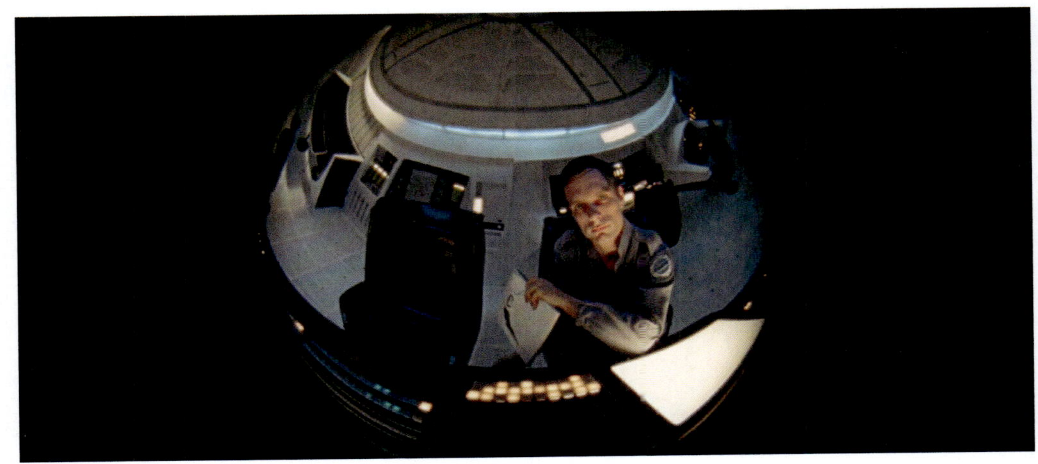

8. Hal lures Dave into a conversation and the second "chess game" begins.

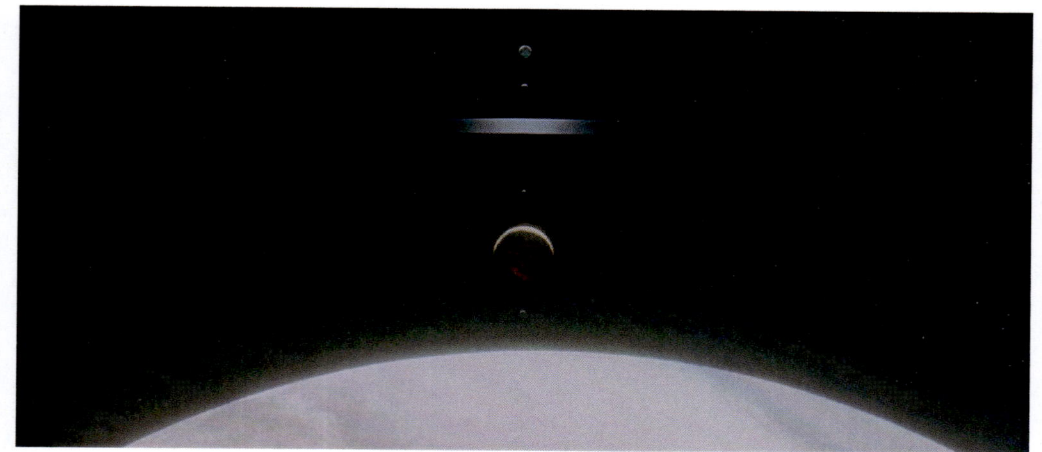

9. Jupiter's moons align with the monolith.

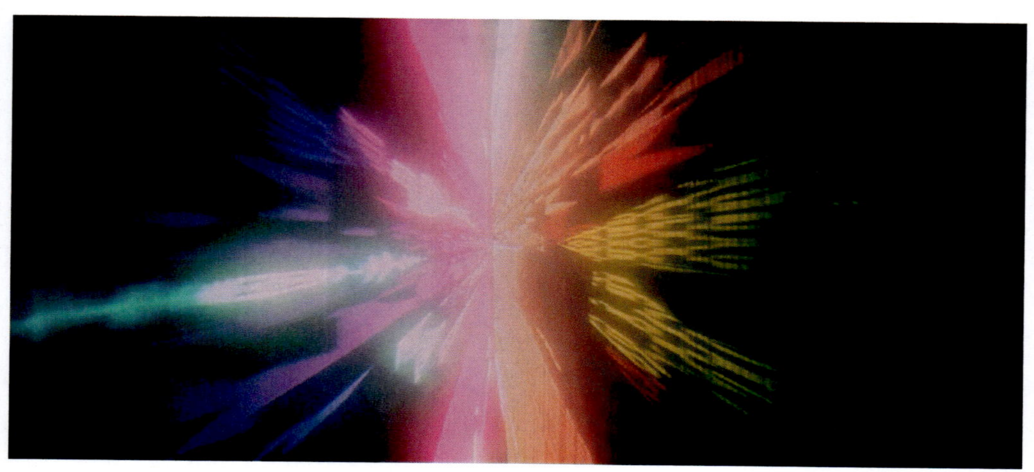

10. A voyage to another galaxy at superluminal speeds presents us with sights unlike anything we have ever seen.

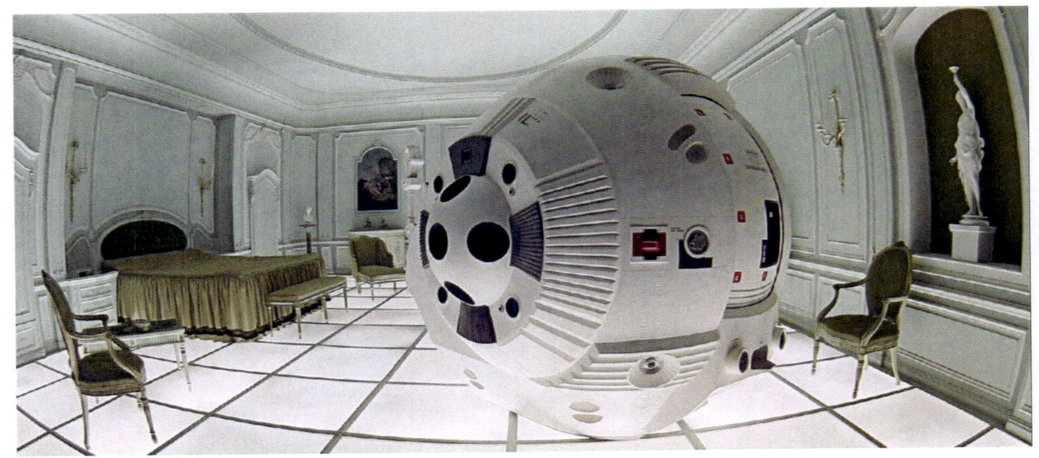

11. Seams in the floor of the alien suite that should be straight appear curved, indicating we have entered a reality different than that on Earth.

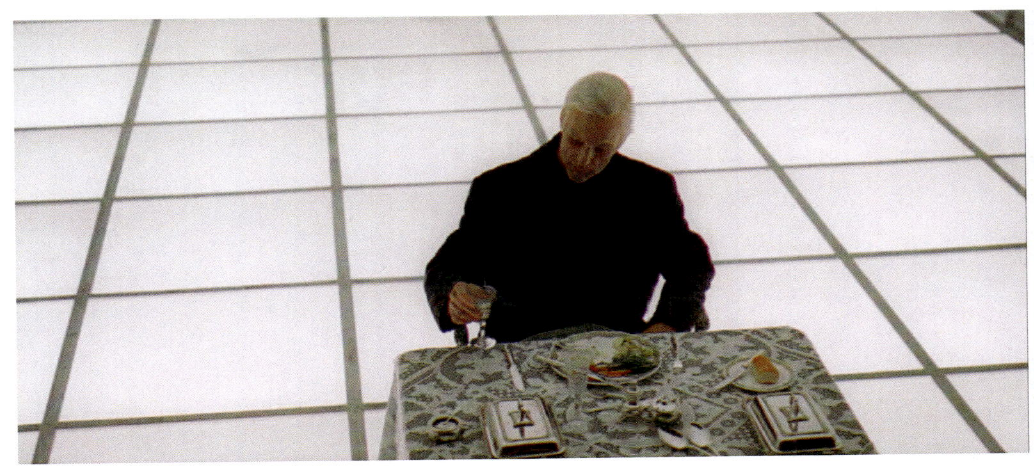

12. Dave Bowman's elegant last supper

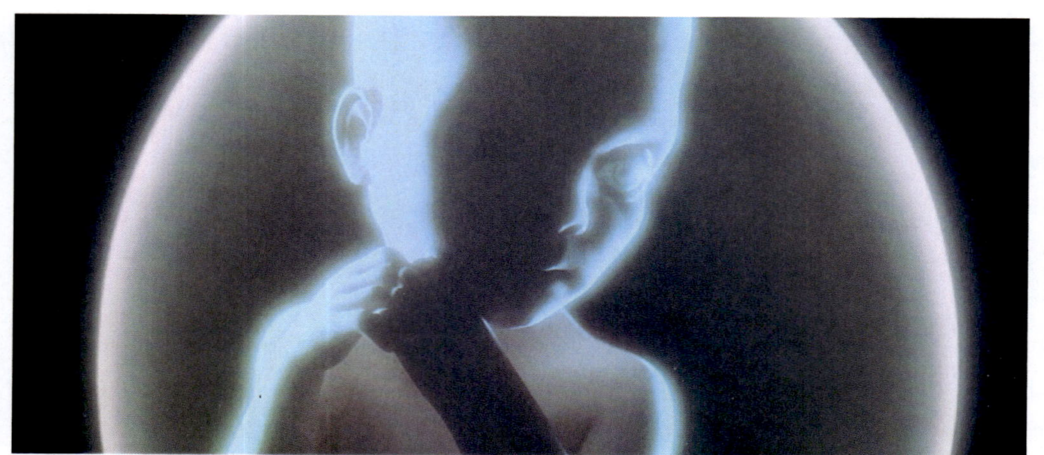

13. The Star-Child gazes down on the earth.

Appendix A
Film Credits for *2001: A Space Odyssey*

Director and Producer: Stanley Kubrick

Screenwriters: Stanley Kubrick and Arthur C. Clarke

Director of Photography: Geoffrey Unsworth

Special Photographic Effects: Wally Veevers, Douglas Trumbull, Con Pederson, and Tom Howard

Editor: Ray Lovejoy

Production Design: Tony Masters, Harry Lang, and Ernest Archer

Art Direction: John Hoesli

Sound: Winston Ryder

Costumes: Hardy Amies

Makeup: Stuart Freeborn

Cast

Dave Bowman	. . .	Kier Dullea
Frank Poole	. . .	Gary Lockwood
Heywood Floyd	. . .	William Sylvester
Moonwatcher	. . .	Daniel Richter
Hal's Voice	. . .	Douglas Rain
Andrei Smyslov	. . .	Leonard Rossiter
Elena	. . .	Margaret Tyzac
Ralph Halvorsen	. . .	Robert Beatty
Bill Michaels	. . .	Sean Sullivan
Floyd's Daughter *(Squirt)*	. . .	Vivian Kubrick

Appendix B

Production Facts about *2001: A Space Odyssey*

Production Company: Metro-Goldwyn-Mayer

Production Location: MGM British Studios, Ltd., Borehamwood, England

Date Production Began: December 29, 1965

Original Release Format: Cinerama Super Panavision 70

Color Process: Technicolor and Metrocolor

Date of Release: April 2, 1968

Date of Release for Final Edited Version: April 9, 1968

Academy Awards Received: 1968 Oscar for Special Visual Effects Kubrick was one of the nominees for the Oscar for Best Director

British Academy Awards Received:

Geoffrey Unsworth for Best Cinematography

Tony Masters, Harry Lang, Ernie Archer for Best Art Direction

Winston Ryder for Best Soundtrack

Other Awards Received: Italy – The Di Donatello Statue for the Year's Best Film from the West

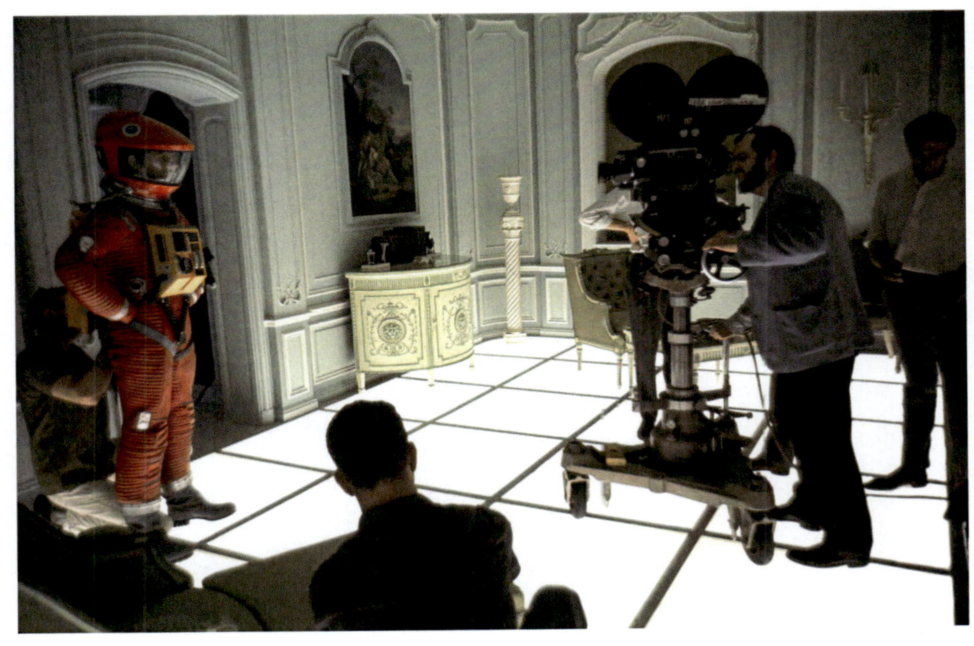

Stanley behind the camera as he and his crew prepare to shoot one of the final scenes in *2001.*

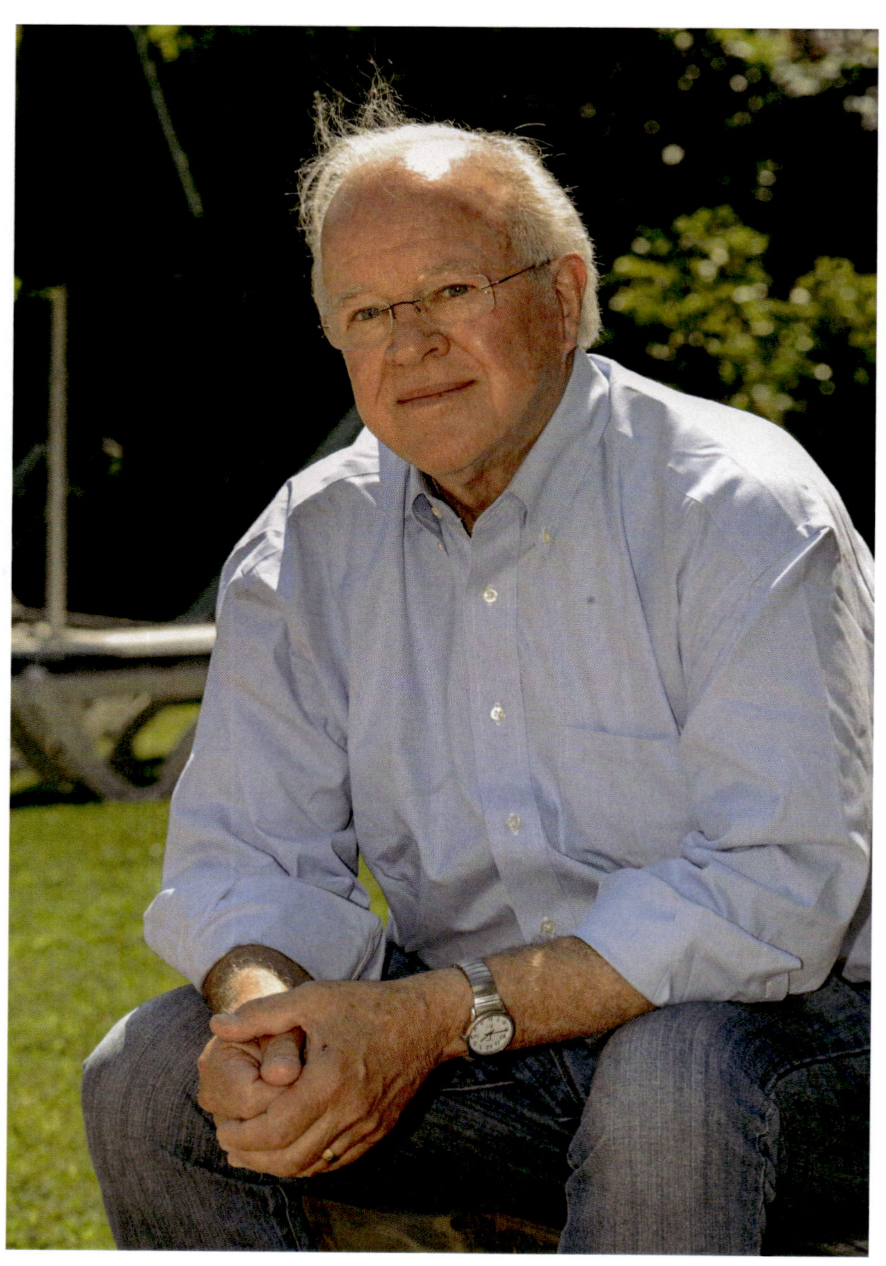

Douglas Trumbull

Appendix C
An Interview with Douglas Trumbull

From his humble beginnings working as an illustrator for a small animation studio, Douglas Trumbull went on to build an illustrious career that has encompassed and continues to encompass a number of ingenious inventions, several films, creative events, and honors. In 1968, Trumbull shared the Academy Award for Best Visual Effects along with co-workers Wally Veevers, Con Pederson, and Tom Howard for his contribution to Kubrick's *2001: A Space Odyssey*. He went on to direct two science fiction features, *Silent Running* and *Brainstorm*. He has also invented and created many special effects that have been used in many films and theme park ride-movies. For several years now he has owned and operated Trumbull Studios where he continues to invent and create his own unique style of movie magic.

On a sunny afternoon in June 2016, I visited Douglas Trumbull in his office at Trumbull Studios in the Berkshires of western Massachusetts. I was hoping to gain more insight into his work, Kubrick's work, and all those who contributed to the production of *2001: A Space Odyssey*. I wasn't disappointed; it turned out to be a fascinating afternoon.

A.H.: What was your background, particularly your technical background, before you started working for Stanley Kubrick on *2001: A Space Odyssey*?

D.T.: My father was an engineer and my mother was a commercial artist. I grew up around both those worlds as a young boy and particularly hung out in my father's workshops of various kinds over the years. As a result, I became extremely comfortable with drill presses, lathes, welders, band saws, and things like that. So my interest and ability in mechanical engineering was built into my genetic code, and my experience and whatever I got from my mother who died when I was seven and was artistic, led me as a young man to believe that I was going to be an architect. I began a pre-architecture curriculum at a community college in Los Angeles and started taking classes in illustration, design, graphic design, mechanical drawing, life drawing, and a broad range of artistic things that were part of a pre-architecture curriculum. But, I was unable to pursue that because my family and I didn't have any money and I had to figure out how to get a job.

During those years my interest in science fiction pervaded all my artwork, so my portfolio was filled with alien planets, space craft, and things like that. I had all my life been interested in Disney animation and I watched every show that Disney made, particularly the *Tomorrowland* programs that featured Werner von Braun and the space program, so I decided to see if there was a way for me to get a job in animation. I had heard through friends of mine that there were barriers to getting a job at Disney because the only entry-level jobs available at the Disney animation department were as an in-betweener[1] and that wasn't what I wanted to do. I was more interested in design and because my portfolio included a lot of spacecraft and planets, I got directed to a company called Graphic Films [founded by former Disney animator Lester Novros] that was making space movies for the Air Force and NASA. That led to more than a year of work at Graphic Films as a background artist on a number of films that included depicting the entire Apollo program. These were films made under contract to NASA and used internally by the government to raise money for the moon program and things like that. This work led to a contract for the movie *To the Moon and Beyond*[2] that would be shown at the New York World's Fair [1964]. I became one of the key illustrators on that movie and did a lot of the artwork and animation for it. Stanley

Kubrick and Arthur Clarke saw *To the Moon and Beyond* at the World's Fair and were looking for a validation of their concept for making this giant-screen, epic science fiction movie which at that time was called *Journey Beyond the Stars. To the Moon and Beyond* impressed them very much and they hired Graphic Films to start doing pre-production illustration for what would later become *2001: A Space Odyssey.* For the first few weeks I worked on a lot of ideas for spacecraft, lunar landing modules, and lunar bases. Then Kubrick decided to move the production to London and terminated the contract with Graphic Films. At that point, I was laid off because there was no other work in the pipeline at Graphic Films, so I asked my boss, Con Pederson, if he could lead me to Kubrick. I was already a big admirer of Kubrick as a director. I had had no interaction with him at all, but I thought *Dr. Strangelove* and other movies that he had made were just stunning motion picture experiences.

A.H.: How did you come to be hired as part of the *2001* crew?

D.T.: I cold-called Stanley Kubrick and said, "I've been working on your movie already. You may not know me, but they've been sending my illustrations to you and I would love to continue working on your film if I could." I don't know the details of any further conversations that took place between my boss, Con Pederson, and Stanley Kubrick, but there must have been some and I immediately got an offer to come and work on the film. Stanley sent plane tickets and hotel reservations for me and my wife to come to London, which we did in a matter of weeks. The indication was that this would be about a nine-month project. So we kind of shut down our lives in L.A., flew to London, met Kubrick, and immediately began working on the film, beginning with animation that was carrying on from what I had done at Graphic Films.

A.H.: When you began working as part of the crew, what were your initial responsibilities and what was your title?

D.T.: I had no initial title. That wasn't even an issue and wasn't up for discussion. I was hired under a tax program called the Eady[3] program, which required that not more than 1% of the production budget be spent on American personnel; 99% had to be British in order for them to qualify for this tax rebate or whatever was going on at the time. So, I was under the radar in the sense that I was one of very few Americans working on the film other than Kubrick. The first task that came up was how to do all these readouts for Hal the computer. We needed simulated computer graphics in the day when computer graphics didn't exist at all. The whole idea of computer graphics was in its infancy. All that was available were very crude vector graphics from Evans and Sutherland,[4] IBM, and other companies that were beginning to emerge at that time.

A.H.: This would have included the readouts or displays you see on the console of the *Orion* as it makes its way to the space station and later inside the *Aries* during its flight to the moon base.

D.T.: Yes, those were all animated films, 16 mm projections from rear projection screens that were built into the sets. There were banks of 16 mm projectors operating behind panels or under consoles. In the *Orion* or *Aries* there might have been two or three, but in the centrifuge or the larger sets on the *Discovery* there could be as many as twelve or sixteen projectors running simultaneously. All those films were shot in 35 mm originally. One of the unexpected aspects of that project was that Kubrick had to be able to direct the actors and the cinematographer and shoot these scenes on kind of an open-ended basis with these graphics running continuously. So, each projector had to be equipped with about a thousand feet of animated footage which had to roll continuously, because if any of those projectors stopped or ran out of film that would be the end of the take. This meant we had to have a lot more animated footage than the length of any take used in the movie.

As I started working on that task, I was using conventional animation techniques which used what we call cue sheets. These are long sheets that would have the description of virtually every frame, the position of every piece of artwork, the color gel required, the designation of the artwork and framing, and everything would have to be laid out. These were traditional animation techniques that we had always used at Graphic Films, but we soon realized that it was going to take ten years if we did it this way because it was so laborious and cumbersome. We needed thousands of feet of animated footage and had to find a way to create it more quickly and efficiently.

At that time, one of the visual effects people hired was Wally Gentleman, who had come from the movie *Universe*[5] that had been made by the National Film Board of Canada. Kubrick admired this movie and had shown it to us and we all admired it. If you get a chance to see it I highly recommend it. It's one of the most beautiful space movies ever made and that was why Wally got hired. And Wally was a very ingenious guy and he said, "Why don't we figure out a different way to do this and I'll help you." So, Wally collaborated with me and Wally Veevers, who became one of the visual effects supervisors on the film and who was also very good at mechanical engineering. We then built an animation stand which is basically a camera aimed straight down on an artwork table where you have cells held on pegs. We decided that we would try to shoot much more quickly, on the fly, without all these laborious cue sheets. We had a camera that Wally found for us that had a variable shutter and an animation motor and we had a sheet of glass with lights behind it so that we could make this backlit artwork for the graphics. I started devising little mechanical devices that connected the motor on the camera to shafts that moved the artwork around automatically on the top of the table. So, instead of having normal peg bars where you would change each cell, we would have one piece of artwork that you could move to different positions on little shafts, pulleys, and gears that were actually being driven by the animation motor. These were made using a mechanical engineering prototyping set that was like a super Erector Set. In America we know it as an Erector Set, that's what I grew up with, but in Germany they had much more sophisticated erector sets that could be used by engineers where they could mock up little devices with pulleys, chains, sprockets and stuff like that. So, it came very naturally for me to engage my mechanical engineering side to solve the problem of creating all these animated readouts. Wally Gentleman then left the production–I'm not certain as to why–and I was left with the problem of continuing to design and shoot all these readouts. We then hired a young animation photographer named Bruce Logan from the animation company in London that we had initially subcontracted the photography to and asked him to come to the studio and work with us so that we would be able to interact immediately. Bruce came onboard and he and I built our own animation stand and concocted this entire process of creating fake computer graphic readouts and doing it very quickly. So, it was this process of artistic design, photography, using color gels, automating with mechanical devices, and animation that enabled us to shoot as much as a thousand feet of animation in one day. This was observed by Kubrick, who is saying, "Ah, these kids are pretty good. They're solving problems for me in a way that's never been done before. Let's give them something more to do."

A.H.: That anticipates my next question: how did you progress to becoming a special effects supervisor?

D.T.: Well, it was a series of steps over time. One of the aspects of my presence during production involves a funny little story. I'm this twenty-three-year old L.A. kid in my Western phase wearing cowboy boots, cowboy hat, and jeans, and all the British guys working on the film thought this was the cutest thing they'd ever seen, and soon came to regard me as sort of a mascot of the crew. At that time the British film industry was highly unionized with very strict division of labor. It was impossible for anyone in say, the engineering department, to work in the props department or for someone in props to do work in the set department and here I was not posing any kind of threat because they couldn't hire any more Americans as this was limited by the Eady program. So, I was able to cross over any of the departmental boundaries in the studio, MGM British Studios at Borehamwood, and they thought this was terribly cute. I was inventing these things and could go to the engineering department and say, "Could you build me this shaft with a sprocket on it attached to a motor," and they'd say, "Sure Doug, we'll have it for you tomorrow." Or I'd go to the wood shop and say, "Could you fabricate this little wooden rig so that I can animate the AE-35 unit I'm working on," and they'd make it for me. So, things like that kept growing and I was able to work anywhere in the studio I wanted without anybody being upset or concerned about it and Kubrick is watching this whole thing. Gradually, our readouts became more and more sophisticated as we were learning more and more as we went along.

One problem that came up was the analysis of the AE-35 unit [part of the rotating antenna on the *Discovery*]. This would tie into Hal's neurosis or whatever you want to call it and I had this idea that we could create this 3-D graphic by building a device that would allow us to animate its rotation.

The AE-35 was actually a gyro system from an aircraft, a very sophisticated-looking piece of mechanical gear that we re-painted to look like a prop. So, I built this device that allowed me to increment the antenna's rotation in hundredths of a degree over a series of frames. Next, I peg-mounted x-ray film, went to London and had an x-ray studio photograph this thing several hundred times at different angles. No one had ever thought of doing this before and Kubrick loved it. We also shot the antenna at an oblique angle which gave it more of a three-dimensional appearance.

A.H.: So you didn't photograph drawings that looked like x-rays, you actually had x-ray photographs taken of your model.

D.T.: Yes, what we really wanted was a three-dimensional x-ray on the screen so you were looking at a three dimensional object that you could see through in color. And that worked really well. We took the hundreds of cells of x-rays and aligned them to each other and they became part of the readouts on the system. Again, Kubrick is watching this and thinking, "Trumbull is doing another weird widget for me and I like it." In the world of computer graphics research at that time, there was this concept called wire-frame animation or vector graphics, but we didn't have them, no one did at that time. They had only been discussed in scientific journals. We needed animation footage of the entire antenna that would rotate and look like a three-dimensional graphic depiction of the antenna rotating. So, I actually built the antenna out of wire, painted the wire white, and mounted it on a rig where I could increment it, photograph it, and animate it. So, we had an animated, rotating antenna in vector graphics at a time when vector graphics didn't really exist. Kubrick really liked this and before long he came to me and said, "How would you like to shoot the moon bus landing?" I said, "Oh, that sounds good. How do you that?" He said, "No, you figure it out." So, he put me charge of the whole landing of the moon bus because I had been working on the moon bus as an artist. The design for the moon bus was created by the studio and had been sent out to be fabricated out of fiberglass by a company that specialized in fiberglass aircraft models. When it came back it looked terrible because it had no detail on its surface whatsoever. I used to do these animation things with an air brush and I could paint right on the surface of the model all these little details, facets, and numbers. Then I developed this process of painting the outside of the model so it would look like it was made of aluminum or titanium and would have different reflective qualities and a lot of fine detail. It was painted with thousands of little friskets which were various-shaped masks made out of celluloid that I would tape to the model and spray paint over with my little airbrush. You would take off one mask and put another one on and keep mixing the paints differently so that eventually you got different refractive and reflective qualities on the surface.

Kubrick was very pleased with that, so he gave me the job of photographing the moon bus landing. He wanted these very bright lens flares that would catch all this turbulence and dust created by the rocket engines and make it look like an exciting place to land. The dust cloud from the landing would also obscure the fact that there was nothing but a very simple cut-out background for the lunar landscape in the distance. The moon bus was actually mounted on a two-inch diameter, black steel rod through which CO_2 gas was flowing to make all the turbulence. So, it was my job to line up the camera, get the lights going, and figure out where to source the lights. I used about fifty 1000-watt projector bulbs, the kind normally used in 16 mm movie projectors. These had a filament with a reflector inside them which created a very directional light source. I mounted them all around the moon bus landing pad so they were all facing towards the camera. These created bright lens flares like we used on the TMA [Tycho Magnetic Anomaly] landing site on the moon, which was shot at Shepperton [now owned by Pinewood Studios]. Kubrick loved the lens flares. Then I engineered these little shock absorbers on the moon bus landing feet to provide a little bounce as it landed. That shot went really well, so Kubrick continued to give me more and more complicated tasks, and this led to me designing lunar landscapes. Kubrick wanted me to take a shot at what the lunar surface would look like. He and I had both been exposed to a lot of the paintings by Chesley Bonestell, the most famous space artist of the day. Bonestell had been painting all of his planets with these rocky mountain peaks that were very jagged. That's what I started doing at first, sculpting jagged lunar mountains out of clay and photographing them to see how it looked. And I said to Kubrick, "I don't think the moon is like this. I've seen photographs of the moon and I think it's going to be very soft, powdery, grey colored, and more softly rounded. A lot of craters are going to define the terrain, but it's not going to be peaky mountains." And he said, "Trumbull I think you're crazy, but show me." So, this became my next challenge. I got a space allocated to me on one of the stages and built this huge lunar landscape out of clay, about twenty by twenty feet, on a big platform, and I sculpted it all by hand. Then I went up about forty feet, near the ceiling and dropped rocks onto this wet clay lunar surface. This created millions of little craters and it looked fantastic, like a real lunar landscape, and I photographed it and showed it to Kubrick. Kubrick said, "Nah, don't like it; we're going with craggy mountains." It wasn't something I would call an altercation, just a difference of opinion. Kubrick said, "I just don't think it's dramatic enough." I said, "Fine, you're the boss and we'll do whatever you want." So he got his way on that one, but ultimately it turned out that I was right. The landscapes I had made for him – sadly, all the photographs of them were lost – actually looked very much like the images that the Apollo astronauts sent back from the moon. Anyhow, that was another one of the stepping stones for me as I took more responsibility for the aesthetic of the lunar landscapes and I made several of them.

Then we worked on another task which involved storyboards of the shots of the lunar bus flying over the lunar landscape. It was my job to figure out how to do the landscapes. The procedure we developed was I would draw a picture of a landscape that I thought would meet Kubrick's design criteria with lots of craters, rills, valleys, mountain peaks, and forced perspective distance things, and I would get that drawing approved by Kubrick for a particular shot. Then I would photograph the drawing in black and white, put the photograph in a projector, and project it onto a big sheet of plywood on the stage. Then we would trace the drawing from the projection onto the plywood. Because we were using forced perspective in order to have depth of field, the plywood couldn't be more than fifteen feet from foreground to background or it would just fall out of focus. That led to me supervising a lot of the lunar landscapes being executed by other sculptors and artists working on the production. So, we were solving all the animation problems, the problem of photographing the miniatures, and how to make all the miniatures look realistically detailed. This led to the *Discovery* and the space station being detailed out in the same way. Eventually, I ended up heading the entire animation department for the movie and we ordered an Oxberry 70 mm animation stand that had a big camera, a table with four peg bars, and back and front lights on it.

The story I'm trying to tell is that there was a lot of engineering behind the artistry that you see on the screen. The Jupiter project was a prime example of this, but I'm telling things out of order because in order to solve that problem we had to invent the slit scan machine. So, I should really begin by telling you how we created the star gate sequence. Kubrick loved using Polaroid cameras during production and had issued Polaroids to the crew so that everybody could take pictures of anything that they were working on that might possibly be used in the production. This way, he would have a two-dimensional image to examine to determine whether or not the aesthetics were right and whether or not the thing might work in the film.

I had a Polaroid camera on my animation stand that I could mount right in front of the animation camera and move up and down. And I had this idea that was based on the work that John Whitney[6] had done on the film *To the Moon and Beyond.* His process involved keeping the shutter of the camera open for a long period of time while other things moved to create what I called a controlled blur. That's what John Whitney had developed for *To the Moon and Beyond* and he went on to make many art movies over the years using that technique. No one at the studio was having any luck with the star gate transition sequence. It was initially designed to be a slot in one of Jupiter's moons, but nobody could figure out how to do that. There were tons of storyboard illustrations, but they either weren't interesting or didn't look good.

I said, "Since the conceptual intention is a transformation or transition forward, I'd like to see if I can come up with something using this idea of streak photography or photographing a controlled blur." By making various mechanical devices, I believed I could control the shape of the blur and create something that was a combination of time and space that might solve our problem of the star gate. I shot a little test using a Polaroid camera mounted on the animation stand and the idea was to create a very thin backlit slit. Everything was black in the room except this slit, which was illuminated from behind, and in front of that I put a piece of artwork that I cranked past the slit. While this was going on, I moved the Polaroid camera from its top position all the way to the bottom in one exposure. This created a plane of light that was a controlled blur, but on the plane was this artwork that had been applied. I realized that we could double this up and create this corridor effect that ultimately became what was called the star gate.

I took the Polaroid photo down to Kubrick's office that minute and said, "Stanley I think this might solve the star gate problem." He said, "Yeah, looks good to me. What do you need?" "Well," I said," I'll need to build this big machine, it's going to be as big as a room, maybe twenty feet long, twenty feet wide, and eight feet high and its going to have all these tracks and motors and I'll have to really scale it up to do it in a large quantity for you." Kubrick says, "Fine, do whatever you need to do." So, I was able to engage Wally Veevers to help me with all the mechanical devices and other departments to do the panes of glass. My animation department built some big backlit worktables to design all the artwork that would be moved behind the slit on the slit scan machine. I wired up the entire machine, soldered all the connectors, and worked out a series of darkroom timers, relays, latches, and switches to make this machine automated so that once it got started, the artwork was loaded, and the camera was loaded it would just sit there and shoot frames automatically for hours at a time.

A.H.: Did you give any specific direction to the artists as to what kind of images or artwork you wanted to photograph?

D.T.: Yes, I worked with them very closely. We shot a lot of little tests and then found and shot thousands of weird patterns from Op art, Zip-A-Tone, textures, microphotographs, and all kinds of stuff that we used to create this abstract light show. As you know, this was in the years of the light show and people tripping out. A lot of Op art was available in books. If you actually analyze the star gate sequence, you would probably be able to identify some of the pages out of art books, if you look closely. This was all photographed and turned into black and white, backlit transparencies with color gels and things. That became the star gate and it solved a really critical part of the movie: this transition through space and time, a concept that had been built into the story.

The slit-scan process worked really well, so the work we had done on it led to the solution of another problem which was that the artists on the film, these were seasoned professional illustrators, were trying to illustrate Saturn, but were having great difficulty and were failing miserably. So Kubrick decided we'll go to Jupiter instead, thinking that it might be an easier planet to paint, but that too proved to be difficult. Nobody could paint a good Jupiter that looked photo-realistic. Then I came up with the idea that we could adapt the slit-scan photographic technique, which had been planar, to a sphere using the same concept. I said, "If I make a flat painting of Jupiter and I build a projector that will project a flat painting onto a very thin strip of paper, rotate that 360 degrees so it's a little semicircle of paper with two images projected on it, I think it will work." The whole gizmo had to rotate 180 degrees to create this sphere of light, which I called a light lathe. I showed Kubrick this idea that I had made from my Meccano set and a little projector that I got from a toy company, did some exposures with my Nikon camera, and proved that we could create a sphere. Kubrick said, "Okay, do that." So that led to a whole other project being built that became known as the Jupiter machine. Again, this involved projectors, optics, motors, and cameras doing eight-inch by ten-inch exposures at f-32 for five hours in the dark. That became our solution for the Jupiter problem. So what I'm getting to is, I kept finding solutions to Kubrick's needs that were a combination of artistry and technology, and that became my art form over the years and what I brought to all the movies I subsequently worked on.

A.H.: Gradually, Kubrick saw that you were helping him realize the vision he had, which was to be something very different than anything that had been put on the screen before with regard to space travel and space exploration.

D.T.: Yes, so that informed me a lot as, ultimately, I wanted to direct my own movies. I came at it from a direction that was similar to Kubrick's, doing research and development which allowed guys like me to solve problems for him. And my realization, because of my unique access to Kubrick and my role in making his movie, was that we were taking the cinema art form to another level of experiential immersion. It wasn't just those things we were doing; it was the fact that this was a Cinerama movie with giant, 90-foot-wide curved screens and six channels of stereophonic sound–an immersive experience. Kubrick, as director, was realizing that he could create something like a flight simulator and make the audience feel like they were really in space and going on this trip. That's when he started consciously deleting all the conventional reverse angle or reaction shots of Keir Dullea or any character explaining, "Oh, it's a star gate. Where did that come from?" He didn't do any of the stuff that normal melodrama would do, whether it was driven by dialogue or character conflict. He said, "I'm over that. We're going to make a movie that is a pure spectacle experience."

A.H.: *2001* was shot on 65 mm stock and released in the Cinerama Super Panavision 70 format. What were the advantages this format offered particularly, with regard to the film's visual special effects and Kubrick's visual style?

D.T.: As you would remember, that was in the day of the road show movies released in these huge theaters with hundreds of seats and gigantic curved Cinerama screens. The Cinerama process had enjoyed considerable success in theaters through the years, but it had some serious problems. The original process used this triple-camera, triple-projector system. MGM had made *How the West Was Won*[7] and *The Wonderful World of the Brothers Grimm*[8] in what was called the three-strip Cinerama process, but it was so cumbersome, and the cameras were so heavy, that it created difficulties for the directors and everybody working on the film. Finally, they decided that even though it was hugely profitable at the exhibition end, it was just too much of a pain in the neck and that's when they switched over to the single 70 mm-camera, Super Panavision system, which made it much easier and more conventional for the directors and actors to deal with. So, here we were with *2001,* one of the first 70 mm wide-screen movies, in the context of an industry that had the desire and the opportunity to create these epic spectacles. And at the same time, the expectation of the audience was that they were going to go to the theater and be engulfed in a big spectacle. Whether it was *Ben Hur,*[9] *Lawrence of Arabia,*[10] or *The Sound of Music,*[11] they wanted a big-screen, epic experience. Kubrick told us that he felt he had a responsibility as a filmmaker to deliver an epic spectacle, but with *2001* it would be in the context of a science fiction movie and this had never been done before. *2001: A Space Odyssey* wouldn't be just a regular 35 mm melodrama like *War of the Worlds*[12] or something; this was an epic.

That led to many of the unique qualities of the movie that Kubrick recognized as it developed that were not built into the original screenplay, nor were they on Kubrick's mind when he got the green light to make the movie. These qualities developed in the course of making the movie. That's a very important component of the equation to acknowledge. That is, if you took the script for *2001* or even if you took the dialogue and descriptions of events in the finished movie and retroactively wrote a screenplay, you wouldn't get past the front door of any major studio today. They wouldn't understand it. They'd say, "Where's the conflict? Where's the drama? Where's this character's development? Where's the backstory?" You wouldn't get to square one with the screenplay for *2001* in the industry today because they never adapted to these giant-screen epics. And the tragic event that happened after *2001*'s release was the multiplexing process, where all these great, wide-screen movie theaters were chopped up into small-screen theaters for economic reasons. This led to the merging of movies with television as one medium, which is where we are today.

A.H.: To the detriment of film?

D.T.: Yes. So, the palate for the epic thing is gone.

A.H.: Would you comment on some of Kubrick's most innovative ideas with regard to the style of *2001: A Space Odyssey*, beginning with cinematography?

D.T.: One of the principal innovations I felt was that Kubrick, in his realization of the epic spectacle of this thing, used extreme wide-angle lenses. In fact, some of these were fish-eye[13] lenses which were twelve inches in diameter. These had been previously developed by Todd-AO and been used on *Oklahoma*[14] and other movies very successfully. With his desire to make you feel like you were actually in the set or in space, Kubrick wanted to use wide-angle lenses most of the time. This meant that there was no place to hide lights, microphones and booms or all of the normal conventions of movie-making. That led to the design under Tony Masters, the production designer, and Jeff Unsworth, the cinematographer, of sets that had the lighting built into them. This is a very unusual attribute of the movie that is very little talked about. For instance, if you were in the centrifuge set, you could shoot that set 360 degrees, move the camera all around, and you didn't have to set any key lights or fill lights or microphone booms. You were there, and almost all the shots that were done in the centrifuge used these fish-eye lenses, including the shot where Gary Lockwood is running around in there as if he were in a squirrel cage. [The centrifuge set is rotating, but Gary Lockwood, playing Frank Poole, actually runs in place.]

A.H.: The shots from Hal's point of view also used fish-eye lenses.

D.T: Yes, a very wide fish-eye lens was used there. So, the whole idea and Kubrick's underlying intention in using point of view or first-person point of view, whether it was Hal's or the audience's, was to make the audience feel like it was *their* point of view, not Keir Dullea's or Gary Lockwood's point of view. The audience was to be in that set or in that space or location so that they could experience it directly for themselves. That was a paradigm shift in the relationship between the audience and the movie. So that pervaded everything.

A.H.: Any innovations in mise-en-scene come to mind?

D.T.: Well, for instance, a lot of radio microphones had to be used in shots because we didn't have a place to hide mics and booms in many of the sets. They often had to be placed on the actors and this was one of the earliest uses of radio microphones on actors in a film. [I did not pursue this question further with Doug as he had already covered many innovations to mise-en-scene in his previous answers regarding models of the spacecraft, displays, interiors, and landscapes that had to be created.]

A.H.: Any innovations with regard to editing?

D.T.: Yeah, it was the avoidance of conventional melodramatic editing which you see all the time in any movie or television show, which is dominated by the over-the-shoulder shot. This is where the eye line to the actor is almost never to camera, it's always off-camera to the other person, but you see the other person over the shoulder. It is one of the most basic conventions in cinema, but Kubrick felt strongly that in this medium it took the audience out of the movie. Kubrick wanted the audience to *become* that other person. So he talked openly to me and others a lot about breaking through the conventions of melodramatic moviemaking using longer shots and lingering longer on them with less editorial intrusion to the scene to let the audience feel like they were there. So that pervaded all the editing and the pacing of the editing.

Then another element developed later as a result of one of the technical limitations we faced. Wally Veevers was the head of a component of the special effects photography department and he built special camera gear for the cinematography. One of the limitations was that when the camera moved, it moved continuously at one speed through the entire shot. It couldn't slow down or speed up during the shot or do a normal live-action kind of movement; it had to move at a constant speed. As Kubrick started cutting all his footage together he realized he had this kind of ballet with these shots of continuous movement. Now, Kubrick had already commissioned Alex North to compose an original music soundtrack and it was done under very arduous circumstances. North became quite ill and was leading the orchestra from a stretcher and so forth. But, Kubrick didn't like the music because it had the same melodramatic problem that he had been having with everything. And that's when he decided to use *The Blue Danube,* which he happened to find in his children's record library, and later the Ligeti pieces, which he discovered totally by accident. His wife was listening to the radio one day and they happened to be playing Ligeti and she called and said, "Stanley I think I've got a piece of music for you that may work." The Ligeti pieces he selected also contained these long continuous strands of music that worked well with his shots. That was part of his whole creative process of these things relating to one another, the limitations of camera mechanics relating to music, relating to editing, relating to first-person point of view. It is part of the whole breakthrough that *2001* and Kubrick made.

A.H.: Could you see that Kubrick was choreographing scenes of these inanimate objects to the classical music pieces he had selected?

D.T.: Yes, but the soundtrack came after everything was shot. One of the aspects of the approach to the making of *2001* that was totally different from anything shot today was that every one of those shots was animated and completed in full length. Then Kubrick would just use the part of it that he liked and could match to the soundtrack. That's why the shooting ratio was 200 to one. That would never be allowed today because visual effects shots are very expensive. Today a visual effects shot would be contracted to a computer graphics company with maybe a two-frame handle on each end. There would be no extra "before" or "after" footage. But on *2001*, all of the visual effects shots were shot as if they were full live-action scenes, a very demanding task. This left Kubrick with all the editorial freedom he might want in post-production to find music or to cut footage to the musical beat or tempo.

A.H.: He made sure he had plenty of raw material to work with.

D.T.: He had tons of raw material that he could then cut together to fill out a stanza or section of a musical piece that he found as stock music. The soundtrack is made entirely of stock music he selected that had been previously recorded by major orchestras.

A.H.: Finally, could you tell us about any of Kubrick's innovations with regard to using sound, and I include in that its complement, his use of silence in the film.

D.T.: Well, Kubrick and Clarke and several other people on the crew had a strong background in science and they knew that there was no sound in space and understood that sound couldn't travel in the vacuum of space. Though that's not one hundred percent true, they believed that it was true at the time, and Kubrick had the nerve to just cut the sound off when you went out into space and then added the only sound that he thought could be there, the breathing over the intercom. I thought that was brilliant. It's not like the scene was entirely without sound; it was a different approach to sound, a kind of first-person sound. The sound editor on the movie was a brilliant guy named Win Ryder who recorded and created a lot of the special sounds for the movie that were quite different and unusual compared to anything that had been done in science fiction movies before. One of my favorite science fiction movies was *Forbidden Planet*,[15] where they developed a lot of electronic sounds and sci-fi sound effects for the soundtrack.

A.H.: That was the work of Bebe and Louis Barron, a husband-and-wife composing team.

D.T.: Right, exactly, and I loved that. That was what was right for that movie, absolutely perfect, but that wasn't what Kubrick wanted. He preferred something completely different that would be much more plausible and more scientifically accurate; and I think that's what he got.

A.H.: At any time, did Kubrick ever mention the films of Norman McLaren, Mary Ellen Bute, Oscar Fischinger, or any other experimental or abstract filmmakers?

D.T.: Not to me, but I think he did discuss them with others. Norman McLaren would have come up in conversation. I know John Whitney's slit-photography work in *To the Moon and Beyond* would have been discussed. Also, the entire crew that worked on *Universe*–Roman Kroitor, Colin Low, and a bunch of the Canadian artists were much discussed as avant-garde filmmakers. As for filmmakers who were doing abstract films at the time, other than John Whitney, I don't remember any conversations, but I would assume Kubrick knew about their work, and certainly thought about it.

A.H.: What was Kubrick like to work with as a director?

D.T.: I thought he was an absolutely fantastic, phenomenal person to work for. I had one of the best professional experiences in my whole career working for Kubrick because he allowed me the freedom to break new ground and think outside the box. He consciously pursued this every day and every week, encouraging everybody to do things differently, to get out of old patterns, to consciously recognize that if everybody was doing things a certain way they should force themselves to think about doing the exact opposite. He would consciously say that and that's what I learned from him.

Then there was this other layer of his personality which was committed to extreme quality control. No flaw was allowed. If there was one fleck in one frame you had to do it again and that annoyed everybody. The seasoned professionals who had made hundreds of movies before *2001* were very annoyed by his persnickety attitude about that. The set construction people would say, "Stanley, what part of the set really needs to look good because the rest will be in the background?" He would say, "No, it all has to look good because I don't know where I'm going to shoot and I may decide to turn the camera around and shoot the back of the set, so make it all look perfect." That drove all those guys crazy. To me, I'm young and had never made a movie before, it all seemed normal and fine. I didn't take exception to it at all. So, it was Kubrick's demanding nature that rubbed a lot of people the wrong way and created stress in people's lives and careers. They had never worked on a movie that took three years to make.

A.H.: Did you feel his striving for perfection was a shortcoming he had as a director?

D.T.: No, I didn't see it that way. He was just different, and I had no problem adapting to it. I had had no previous experience anyway, so the fact that my nine-month stint turned into two and a half years for me was challenging and unexpected, but it was definitely the opportunity of a lifetime and I loved every minute of it. So that's just a difficult thing that a lot people saw in Kubrick, his being neurotic, demanding, or pedantic. I don't know, but a lot of people had bad things to say about Stanley. I never had anything bad to say about Stanley except that I was ready to move on and apply what I had learned to my own career. Over the years, I've done presentations on the making of *2001: A Space Odyssey* and talked about *2001* many times and I've always spoken respectfully about Stanley Kubrick as one of the greatest filmmakers of all time, in my opinion. Stanley Kubrick's daughter came up to me after one of these things and said, "You are the only person I've ever heard talk about Stanley in a positive way."

A.H.: Vivian Kubrick?

D.T.: Yes, Vivian and Anya. Actually, it was Anya that said that to me. She has passed away now, but I thought that was nice. I truly thought the guy was amazing and he never got an Academy Award as best director or anything. He was kind of the way I am. He was an outsider, an outlier, not a conventional, play-by-the-rules director. Every one of the movies he made was completely different than every other movie he had made. He would spend at least four years on each movie, researching it, studying it, and developing it. It tock a huge amount of patience for studios to deal with him and a huge amount of patience for actors to work with him, because he wanted so many takes. So, he had a reputation that was peppered with people's reactions to his style. And I say, if you want to work with Stanley Kubrick, that's what you get, and you are going to come out a better person for it. So, go along for the ride, and I think many people did.

A.H.: You admired his style of managing people and working with people.

D.T.: Yeah, he was always courteous, always thoughtful, never raised his voice on the set. He made every effort to make people comfortable, but that was still in the context of being very demanding. He didn't smooth-talk you if you made a mistake. You had to go back and do it again.

A.H.: What would you say were his greatest strengths as a director?

D.T.: I think it was his almost genius-level understanding of the technology, the craft, and art that had to be in play for that particular movie. I never worked with him on any other movie, but I just know his level of attention to detail was extremely refined, thoughtful, amazing, and quite different from anything I experienced with any other directors I've worked with, who were much looser, easier to get along with, and more flexible. I learned a lot working with Stanley and it stuck with me to this day. That's why I'm here and not out in L.A. We are doing everything differently. I am still a big advocate of his attitude and style of direction because we're still thinking outside the box. He would love what we are doing here. In fact, what we're doing here right now with this new process–changing the frame rate, changing the aspect ratio, and changing the size of the theater–runs very much against the grain of Hollywood.

A.H.: What would you say were his most significant weaknesses as a director?

D.T.: Well… I'm reluctant to criticize him in any way because I don't really feel that way. I would just say that I had experiences with him that revealed that he had tremendous frustration with a lot of his human interactions. He did his best to try to adapt to other people's shortcomings or inability to understand what he wanted. And so, as I said, he never, that I know of, raised his voice or got angry on the set. He would do everything he could to help everybody get to the next take or make happen whatever was needed to move the project forward.

A.H.: Could you share any stories with us that would give insight into Kubrick the director or the creative side of his personality?

D.T.: I think one of the most notable aspects of his personality and creative process was that he lived inside of his movies. His life was not a personal life, it was a movie life. He was immersed in the making of that particular movie twenty-four hours a day, seven days a week. The movie pervaded everything, and it was very difficult for Stanley to take a break, have a party, go out to dinner, or do any of the normal social things human beings like to do. So, it was very frustrating for him that the crew would take the weekends off. You could tell on Monday morning that Stanley had spent all weekend trying to figure out ways to do things better and how to organize the production of the movie better. And he was very frustrated with the fact that human beings, including all of us on the crew, everybody at the studio and the contractors that were hired, were not up to his expectations of in-depth communications, efficiency, rapidity, and performance. He was just constantly a frustrated human being because his intellect was on an IQ level that was way above everybody's heads. This was a very big frustration that he lived with, so every Monday morning he would come in with some new plan. He said, "This week everybody is going to have to dictate what they want to say to me into a little tape machine and submit it to my secretary, so she can distill it for me because I don't want to have to listen to what you have to say. I want her to distill it, so she can give it to me all at once, so I can handle it." That lasted about a week. He bought everybody little pocket recorders, but it became a big joke. It didn't work and socially none of us were interested in it; we just made fun of it. That led to crazy things at the dailies every day where we were usually screening shots involving special effects because the actors were long gone. We knew what Stanley was going say and we knew what we would say, and we would pre-record our responses so, in the dark, when Stanley would say, "What's wrong with that?" Someone would already have a pre-recorded response that would say, "Well Stanley, we've screwed up on frame number 52B, we know it and we're going to re-shoot it tomorrow." It was our way of having fun with his frustration, but for him it was probably really hard because we drove him crazy.

We even staged a murder in the screening room where I had a fake argument with the animation cameraman. We had a starting pistol that was quite loud, and we had the cameraman murder me in the screening room over some mistake I had made which we knew Stanley was going to criticize. He would know it was funny and he would tolerate the fact that we were doing crazy stuff. Then the next Monday would come and he says, "Now I want everyone to write down their notes on three-by-five cards. We're going to get organized and put them up on a bulletin board." We all tried to do that. Then the next week he says, "I found this guy who's got this filing system using five by seven cardboard cards that have perforated holes in them." The perforations are based on certain criteria that the card has, like a card file in a library. Rods could be put through the stack which would only pull up cards that were related to some particular issue. Stanley said, "We're going to start a whole new program here to keep things organized using these cards." That lasted about two weeks.

One of the funniest ones was when he came in one Monday and said, "From now on I want everyone to speak to me in pigeon English." We said, "What does that mean?" And he said, "I want you to say exactly what's on your mind without saying anything that is superfluous. For example, on Friday, Joseph from the paint department came up to me when we were quite busy on the set and said, [Doug affects a British accent] 'Stanley, remember when we were over on stage 32; you had said something to the effect that the color of the seating arrangement on the vinyl was to be something with a little bit more blue. And we have a problem in the procurement department as we don't have any more blue and we need to acquire some, blah, blah, blah.' You know what? Come up to me and say what you want: want blue paint. Three words, that's all I want to hear." So that entire week that was the way you had to talk to Stanley, want blue paint, will be late today. No extra words, he just wanted you to get directly to the point. So, it was this kind of dynamic that was funny, charming, frustrating, and infuriated a lot of people.

A.H.: What do you believe were Arthur Clarke's most important contributions to the film?

D.T.: To stay out of the way. I mean, I love Arthur. I think he's a brilliant guy and early in the production, he and Kubrick were a kind of team developing the screenplay and collaborating on the project. Kubrick had optioned the rights to Clarke's story called "The Sentinel," which was about finding an artifact on the moon which would be this sort of intergalactic alarm clock or warning system set in motion by aliens millions of years ago. This was a very amazing idea that Kubrick had purchased from Clarke. As things developed, Clarke was contributing to the screenplay and writing scenes and sequences, but Kubrick didn't like what he was writing at all. Now, I can't tell this story as well as others because I wasn't involved in it directly, but the basic issue was that they came to a gentleman's agreement that a movie is different from literature. Arthur is a wonderful writer with a lot of flowery descriptions and so forth, but the written word doesn't translate necessarily into making a movie. So Kubrick said, "Arthur, just go away; I'm making a movie, you want to write your book. I don't care if they are different. When we finish you can have a book that's different from my movie. It's fine." They had an agreement to disagree. Consequently, Arthur was almost never around during the making of the movie and I had virtually no interaction with him at all.

A.H.: What were some of the things you learned working on *2001* that proved to be valuable during the making of *Silent Running*?

D.T.: Oh, that's a good question. I've never been asked that question. I would say that working on *2001* gave me a kind of fearlessness. I was unafraid of any aspect of making a movie even though I had never made a movie before, or written a script, or directed actors. In the process of making *2001,* I had sort of unconsciously recognized that the component of directing a movie that involves directing actors and actresses was the easy part. While the challenging and fun part that involves solving technical problems and doing stuff no one's ever seen before, instead of the same old same old, is the hard part. So, I came out of *2001* completely fearless, feeling like I could make my own movie, write my own script, start my own company, and that any movie that came along after *2001* was going to be a piece of cake. So, I wrote the original treatment for *Silent Running*[16] because I was into science fiction and had my own ideas about alien contact. I also had my own feelings about humanity and I didn't entirely agree with Kubrick's idea that the future was going to lead to these automaton astronauts who just took orders and had no emotional reaction, disposition, or conflict of any kind. I thought it would be fun for me to bring my own kind of humanity to a movie that would be the opposite of *2001* emotionally in many respects, but be true to science fiction, technology, and space. I had to bring these ideas together in a little sci-fi movie that had to be, by nature, very inexpensive. I knew I could design *Silent Running* in a way that I could do everything that had been done on *2001*, but for a tiny fraction of the cost. This was because I no longer had to do research and development. I knew how to do it all.

For example, one of the biggest challenges on *2001* was making Saturn; everybody gave up on Saturn. Well, by the time *2001* was over, I could do Saturn in ten minutes. So, I put Saturn into *Silent Running*. It was just like falling off a log for me to direct my own movie. As I began writing *Silent Running*, I intended to have as small a cast as possible because I had no training or experience in directing actors. When I selected Bruce Dern,[17] I picked him because I met him and liked him and I trusted him. I found in the process of making *Silent Running* that he helped me learn how to direct actors. Bruce Dern taught me what he was doing. He said, "I'm a method actor and here's what I'm going to do: you want me to cry on camera, I'll cry on camera and say the words that are in your script, but emotionally I'm going to be reliving the death of my daughter." I said, "Wow, that's heavy." But, by that time I was fearless and I said, "Okay, you want to do that, I'm going to be with you. We'll get the camera rolling and we'll shoot it." I learned a lot about directing actors working with Bruce Dern and I found out this really isn't a mysterious process. The hard part is still doing the visual effects and production design.

The same thing proved true for directing the camera setups. I had no clue how to direct. I'd never really thought about it consciously, you know, master shots, over-the-shoulder shots, and all the conventions of movie directing. But, the crew I'd put together for *Silent Running* taught me how to direct and I learned it in a few days. I realized this is not really the hard part. People tend to think it's the hard part, but the hard part is being innovative, doing something people haven't seen before. If you just want to make another episode of a television series or a sequel to a movie and it's the same as the previous movie, you're not innovating, you're not doing anything difficult, and it's just the same everyday thing. I've never felt that way and I always supported Kubrick's willingness to think outside the box.

A.H.: In a broad sense, what do you believe Kubrick was trying to achieve with *2001: A Space Odyssey*?

D.T.: I feel that he wanted to legitimize intelligent, scientific thought about the universe, God, life, and mankind and do it in a way that was totally different from any kind of pedantic, religious belief system or anything. But, I think his biggest contribution was bringing this concept of where does man fit in the universe to the surface in a way that allowed everybody watching the movie to interpret it in their own way. He didn't bring to it any locked-in predisposition as to whether there was a Jesus Christ or not, whether there was a God or not, or anything. He wanted to create an experience where people could contemplate the enormous and scary immensity of the universe and time and space. That's why I rate it as one of the most important movies ever made.

A.H.: How has *2001: A Space Odyssey* influenced other science fiction films?

D.T.: I think that *2001* unleashed a kind of comfort zone that allowed filmmakers to believe that they could actually make a movie about space or science fiction and not be stuck with conventions of being on the ground or being on the earth. It unleashed this whole idea that had been true for science fiction writers for centuries, which was that science fiction as a literary or cinematic art form could be taken more seriously. So you had film directors like George Lucas who came along with *Star Wars,*[18] and without *2001* breaking the new ground that it did, *Star Wars* probably never would have happened. There is a lot in *Star Wars* production design-wise and style-wise that is very similar to *2001,* derivative in some ways. I don't mean to be derogatory because I have great admiration for George Lucas. He brought a whole other dimension to science fiction production that really grabbed everybody's attention and legitimized the idea of alien contact and a universe filled with the same things that happen here on earth. I mean, we have wars amongst ourselves and the universe has wars amongst itself, there's good and bad out there just as there is good and bad here, and there are epic issues about conflict and heroism that George was quite serious about.

I think *2001* unleashed a whole new genre or continent of content that became as ubiquitous as westerns or film noir had been. *2001: A Space Odyssey* legitimized the whole world of science fiction production and thought about things that are bigger than murder, war, love, or the conventions of movies–mysteries.

End Notes for Interview

1 An in-betweener is a person who draws in-between drawings necessary for character animation and, sometimes, non-character animation, in an animation studio. An in-betweener usually works for an assistant animator and the position is regarded as an entry-level position in most animation studios.

2 *To the Moon and Beyond* was a short, futuristic science film created by the animation staff at Graphic Films which included Con Pederson, Douglas Trumbull, and John Whitney among others. The film was completed in 1963 in Cinerama 360 and was exhibited at the 1964 World's Fair held in New York.

3 The Eady Levy, named after Sir William Eady, was a program that obtained revenue from box office receipts in the United Kingdom. Funds derived from this tax were used to support studios and film production companies operating in Britain and allowed for tax breaks on their income. Passed by Parliament in 1950, the Eady Levy was also known as the British Film Production Fund and was administered by the British Film Fund Agency with the intention of increasing employment of British actors and film crew people in the United Kingdom.

4 Ivan Sutherland (b. 1938) is a renowned computer scientist and inventor who did pioneering work in the field of computer driven cathode ray tube displays during the early 1960s which later evolved into computer graphics. He invented the "Sketchpad" program in 1962 and continued to work on computer graphics and virtual reality displays during the late 1960s. In 1968 he co-founded Evans & Sutherland with colleague David Evans and continued with research and development on accelerated 3-D computer graphics.

5 *Universe* was a short, nonfiction narrative film that contained both live action and animated footage. It was directed by Roman Kroitor and Colin Low and released in 1960 by the National Film Board of Canada. The soundtrack included music composed by Eldon Rathburn and a voice-over narration by Douglas Rain. In 1961 it was nominated for the Academy Award for Best Documentary Short Subject.

6 John Whitney (1917-1995) was an American animator, composer and inventor. He was interested in astronomy, space exploration,

music, composing, and mechanical devices related to drawing and animating abstract forms. Whitney collaborated with his brother James on several abstract films and also collaborated with designer Saul Bass on the title animation for Hitchcock's *Vertigo.* In 1960 he founded Motion Graphics, Inc.

7 *How the West Was Won,* directed by John Ford, Henry Hathaway, and George Marshall, was released in 1962 in Europe and in 1963 in the US, in Cinerama.

8 *The Wonderful World of The Brothers Grimm,* directed by Henry Levin, was released in 1962 in Cinerama.

9 *Ben Hur,* directed by William Wyler, was released in 1959 in 70 mm, MGM Camera 65.

10 *Lawrence of Arabia,* directed by David Lean, was released in 1962 in Cinerama Super Panavision 70.

11 *The Sound of Music,* directed by Robert Wise, was released in 1965 in 70 mm Todd-AO.

12 *War of the Worlds,* directed by Byron Haskin, was released in 1953 in 35 mm Technicolor.

13 Fish-eye lenses have very short focal lengths that capture an image of the subject at an extremely wide angle. Objects photographed with a fish-eye lens will curve around the center of the frame and appear quite distorted.

14 *Oklahoma,* directed by Fred Zinneman, was released in 1955 in 70 mm Todd-AO.

15 *Forbidden Planet,* directed by Fred Wilcox, was released in 1956, in 35 mm Eastman Color. Bebe and Louis Barron composed an all-electronic music soundtrack for the film and were credited for the "electronic tonalities" they created.

16 *Silent Running,* directed by Douglas Trumbull, was released in 1972, in 35 mm Technicolor, wide-screen, anamorphic format.

17 Dern plays a botanist named Freeman Lowell who is the protagonist in the film's story.

18 *Star Wars,* directed by George Lucas, was released in 1977 in 35 mm Technicolor Cinemascope.

Appendix D

The Cinerama Super Panavision 70 Process

Principal photography for *2001: A Space Odyssey* was recorded on 65 mm color negative film stock which, after editing, optical effects, addition of the soundtrack, and other post-production processes, was used to produce a 70 mm release print for distribution to theaters to be projected in a format known as Cinerama Super Panavision 70.

The Cinerama Super Panavision 70 process evolved from a wide-screen format invented in 1952 and known as Cinerama. The original Cinerama process involved shooting with three cameras during production and projecting three release prints, using three synchronized projectors, onto a deeply curved screen in specially constructed theaters. Though Cinerama offered an exciting, immersive experience and quickly became popular with audiences, there were a number of technical difficulties that arose both during the production (shooting) phase of the film and during theatrical exhibition that discouraged directors and studios from using it.

Efforts were made by several of the major studios along with the camera manufacturer, Panavision, to develop an improved 70 mm process that would be easier to work with and exhibit. By 1965, Cinerama had been replaced by three new 70 mm processes: Ultra Panavision 70, Super Panavision 70, and Super Technirama 70. All three required only a single 65 mm film camera and a single projector for exhibition. Super Panavision 70 was projected from a 70 mm release print using spherical lenses onto a curved screen. The projected image had an aspect ratio of 2.21 to 1 and also offered six magnetic tracks of stereophonic sound. Cinerama Super Panavision 70 prints could also be shown in theaters with flat screens, though the surround effect created by the curvature of the Cinerama screen was absent.

An anamorphic 35 mm print of *2001* with a 2.35 to 1 aspect ratio was later prepared and released for theaters not equipped for 70 mm projection.

For a detailed history of wide-screen formats such as Cinerama and Cinerama Super Panavision 70, see John Belton's *Widescreen Cinema* (Cambridge: Harvard University Press, 1992), p. 85.

End Notes

1 C[8va] represents a note one octave above the first C in the sequence.

2 Although dictionary definitions of archetype are helpful, to truly understand the concepts behind the meaning of the word, the descriptions and insight offered in the original writings of psychiatrist Carl G. Jung are particularly valuable. Refer to *Collected Works of C. G. Jung, Vol. 9 (Part 1): Archetypes and the Collective Unconscious,* edited and translated by G. Adler and R.F.C. Hull (Princeton: Princeton University Press, 1968).

3 Schneider, Michael S., *A Beginner's Guide to Constructing the Universe* (New York: HarperCollins Publishers, Inc., 1994), p. 39.

4 Schneider, p. 38.

5 The titles used for Acts I and IV are identical to those used in the film. The titles for Acts II and III were added by the author to suggest an interpretation of the act's main events or a possible premise.

6 A graphic match is created during the transition or cut from one shot to another when an object or form appearing in the first frames of a shot matches, at least roughly, the position, shape and/or color of an object or form appearing in the last frames of the preceding shot.

7 One of four major techniques of style that compose a film's stylistic system, mise-en-scene is a French term used in theater to describe how scenes are staged. In film production it includes lighting, set design, character behavior, blocking, properties, costuming, hairstyle, makeup, and any other staging techniques used in front of the lens during production.

8 TMA-1 is an acronym for Tycho Magnetic Anomaly.

9 When discussing plot in a film, I am referring to all of the events both diegetic and non-diegetic that take place on the screen in the order in which they are presented. When discussing plot as defined in literature, I shall refer to it as literary or story plot.

10 The voices, or to be more precise, the vocal utterances heard in this sequence are from a short, music-theater composition by Gyorgi Ligeti titled *Adventures* that did not appear in the film's credits. Roughly three minutes of *Adventures* was apparently

edited, filtered, and remixed to create an otherworldly accompaniment to Bowman's walk through the alien suite.

11 "… the second law of thermodynamics is related to the fact that some processes are irreversible; that is, they go in one direction only. However, there are many irreversible processes that are not easily described by the heat engine or refrigerator statements of the second law. Examples are the free expansion of a gas or a glass falling off a table and shattering when it hits the floor. … There is a thermodynamic function called entropy S that is a measure of the disorder of a system. … In an irreversible process, the entropy of the universe increases." Paul A. Tipler, *Physics for Scientists and Engineers,* Vol. 1, 3rd ed. (New York: Worth Publishers, 1991), p. 576.

12 R. Feynman, R. Leighton, and M. Sands, *The Feynman Lectures on Physics,* Vol. 1 (New York: Addison-Wesley Publishing, 1963) pp. 46-7.

13 "…the entire universe is in a glass of wine, if we look at it closely enough." Feynman, pp. 46-8.

14 I am much indebt to David Bordwell and Kristen Thompson for the concepts and definitions of terms put forth in their excellent text, *Film Art – An Introduction,* 10th ed. (New York: McGraw-Hill, Inc., 2013).

15 Bordwell and Thompson p. 500.

16 Bordwell and Thompson p. 500.

17 I have called these sections acts based on the four divisions in the screenplay originally labeled A, B, C, and D by authors Kubrick and Clarke. Each stands as a unit of dramatic action which takes place at a different time and in a different location.

18 Some writers have referred to the "Dawn of Man" section as a prologue; however, this term may lead to misunderstanding the film's story plot. I prefer to call "The Dawn of Man" section the first act of the film. A prologue would usually offer either a voice-over or text which would introduce the story, but these are absent and this section offers much more than a simple introduction or preface to the film's story. The first representatives of the human race including the main character (referred to in the screenplay as Moonwatcher) are introduced along with the adversities presented by their environment. By the end of the first act (the first major plot point) the first sign of extraterrestrial intelligence (the

monolith) has appeared and ape-man has begun his ascent to modern man and invented his first tool so both the dramatic situation and the path of the protagonist have been established.

19 "... 'myth-forming' structural elements must be present in the unconscious psyche. These products are never (or at least very seldom) myths with definite form, but rather mythological components which, because of their typical nature, we can call "'motifs,' 'primordial images,' types or–as I have named them–*archetypes*." Carl G. Jung, *The Collected Works of C. G. Jung*, Vol. 9, Part 1, 2nd ed., (Princeton: Princeton University Press, 1968) p. 152.

20 Joseph Campbell, *The Hero with a Thousand Faces*, (Princeton: Princeton University Press, 1949; reprint ed. New York: MJF Books) p. 245.

21 Carl G. Jung, *The Collected Works of C. G. Jung*, Vol. 9, Part 2, 2nd ed., (Princeton: Princeton University Press, 1959) p. 266.

22 Carl G. Jung, *The Collected Works of C. G. Jung*, Vol. 9, Part 1, 2nd ed., (Princeton: Princeton University Press, 1969) p. 165.

23 The aspect ratio of a rectangular image or frame is the number obtained when the image's width is divided by its height.

24 See Appendix D for more information about the Cinerama process.

25 The term subjective camera refers to shots that have been taken from a character's point of view.

26 Fish-eye lenses have very short focal lengths that capture an image of the subject at an extremely wide angle. Objects photographed with a fish-eye lens will curve around the center of the frame and appear quite distorted. Plate 8, a shot taken from Hal's point of view, provides an excellent example of a shot taken with a fish-eye lens.

27 Refer to plate 11. In this shot the curving as opposed to parallel lines suggest Bowman has entered a world where space and time have been and can be altered. The curving or distortion is created through the use of wide-angle lens cinematography.

28 The soundtrack for *2001: A Space Odyssey* is available on the CD *Original Motion Picture Soundtrack 2001: A Space Odyssey*, produced by Rick Victor and David McLees, Turner Classic Movies Music/ Rhino Movie Music, R2 72562.

29 Non-diegetic refers to any music, sounds, or visual elements within the frame that occur during a film that are not part of the story world, i.e. not sounds or visual components of the shot that the characters in the story could be aware of, for example titles, credits, and soundtrack music that is not sourced. Diegetic, on the other hand, refers to all sounds and visual components of the shot that are part of the story world in a narrative film.

30 In 1961, three programmers working under the supervision of research engineer John Pierce at Bell Telephone Laboratories successfully programmed an IBM 704 computer to sing and provide a simple musical accompaniment to the once popular nineteenth-century love song, "Daisy Bell." This was reputed to have been the first instance of a computer-generated voice singing a song. Pierce was a friend and colleague of Arthur Clarke and when Clarke visited Bell Labs in 1962 Pierce demonstrated his singing computer. Clarke was very impressed with what he heard and several years later suggested to Kubrick that Hal sing this song in *2001*. Interestingly, "Daisy Bell" was written by English songwriter Harry Dacre in 1892 to celebrate the advanced technology of the safety bicycle.

31 In Arthur Clarke's novel, *2001: A Space Odyssey*, the monolith serves as a kind of training machine for the apes and actually issues commands to them which they receive inside their brains. Depending on how well they perform, the monolith rewards them with spasms of either pain or pleasure. Kubrick, on the other hand, preferred the monolith to remain mysterious and left its role far more open to interpretation. He wanted to avoid being obvious and specific about what the monolith was doing.

32 It is important to bear in mind that although Clarke's earlier short story, "The Sentinel," inspired the story for *2001: A Space Odyssey* and was the source of some of the ideas seen in the film, it was never Kubrick's intention that the screenplay be an adaptation of this work. The dialogue and events that occur in the film vary considerably from what was written in "The Sentinel" and earlier versions of the screenplay.

33 C. G. Jung, *Synchronicity – An Acausal Connecting Principle,* translated by R. F. C. Hull, 2010 ed. (Princeton: Princeton University Press, 1973).

34 Victor Mansfield, *Synchronicity, Science, and Soul-Making* (Peru: Open Court Publishing, 1995), p. 22.

35 "Sooner or later nuclear physics and the psychology of the unconscious will draw closer together as both of them, independently of one another and from opposite directions, push forward into transcendental territory, the one with the concept of the atom, the other with that of the archetype. The analogy with physics is not a digression since the symbolical schema itself represents the descent into matter and requires the identity of the outside with the inside. Psyche cannot be totally different from matter, for how otherwise could it move matter? And matter cannot be alien to psyche, for how else could matter produce psyche? Psyche and matter exist in one and the same world, and each partakes of the other, otherwise any reciprocal action would be impossible. If research could only advance far enough, therefore, we should arrive at an ultimate agreement between physical and psychological concepts. Our present attempts may be bold, but I believe they are on the right lines. Mathematics, for instance has more than once proved that its purely logical constructions which transcend all experience subsequently coincided with the behavior of things. This, like the events I call synchronistic, points to a profound harmony between all forms of existence." Carl G. Jung, *The Collected Works of C. G. Jung*, Vol. 9, Part 2, 2nd ed. (Princeton: Princeton University Press, 1969) p. 261.

36 L. von Bertalanffy, "General System Theory", *Main Currents in Modern Thought*, 71, 1955, p. 75.

37 *Encyclopedia Britannica*, ca. 1936, "Holism," by Jan Christian Smuts.

38 Joseph Campbell, *The Hero with A Thousand Faces*, (Princeton: Princeton University Press, 1949; reprint ed. New York: MJF Books), p. 20.

39 Phil Hardy, ed., *Science Fiction*, (New York: William Morrow and Co., Inc., 1984), p. 279.

40 David Bordwell, *Narration in the Fiction Film* (Madison: University of Wisconsin Press, 1985), p. 157.

41 *The Making of Kubrick's 2001*, ed. by Jerome Agel, (New York: Signet/The New American Library, 1970), p. 113.

42 Albert Halstead, "Towards a Theory of Musicography" (unpublished treatise and notes, 1972).

43 John F. Dashiell, gen. ed., *Psychological Monographs*, (Columbus: The Psychological Review Co., 1938), vol. 50 No. 2: *Color-Music* by Theodore Karwoski and Henry Odbert.

44 *To the Moon and Beyond* was a short, futuristic science film created by the animation staff at Graphic Films which included John Whitney, Con Pederson, and Douglas Trumbull. The film was narrated by Rod Serling and completed in 1963 to be exhibited at the 1964 World's Fair held in New York. The final print was released in the 70 mm, Cinerama 360 format to be projected overhead on the inside of a hemispherical screen. Kubrick and Arthur Clarke visited the fair to see this film and greatly admired it.

45 *The Making of Kubrick's 2001*, ed. by Jerome Agel, (New York: Signet/The New American Library, 1970) p. 190.

46 Agel, p. 192.

Index

About the Author

Albert Halstead is founder and director of Leonine Productions, LLC in Waverly, New York, which has been in business since 1975. Halstead graduated from Clarkson College of Technology in 1968 with a Bachelor of Science degree in Chemistry and later attended Ithaca College, where he received a Bachelor of Science degree in Film Production in 1997. Halstead has worked professionally in film production in both the northeastern US and in Los Angeles.

Halstead has written, directed, and produced three experimental films, a documentary, a dance film, a comedy, and six short dramas. His most recent films include *My Cocktail with Hollis*, *My Brother's Keeper*, and *Gothic Nightmare*, which received several awards at independent film festivals. He has also lectured extensively on science fiction films, film noir, film music, film style, the history of artificial intelligence and robots in film, and Stanley Kubrick's work on *2001: A Space Odyssey*.